065989   168.76

KU-746-256

the last

This book is to be returned c
the last date stamped b

ML

2

01

STRATHCLYDE UNIVERSITY LIBRARY

30125 00064032 5

ANDERSONIAN LIBRARY
★
WITHDRAWN
FROM
LIBRARY
STOCK
★
UNIVERSITY OF STRATHCLYDE

Fundamentals of
Optical Fiber
Communications

Academic Press Rapid Manuscript Reproduction

Fundamentals of Optical Fiber Communications

Edited by

Michael K. Barnoski

Hughes Research Laboratories
Malibu, California

Academic Press, Inc.   New York, San Francisco, London, 1976
A Subsidiary of Harcourt Brace Jovanovich, Publishers

COPYRIGHT © 1976, BY ACADEMIC PRESS, INC.
ALL RIGHTS RESERVED.
NO PART OF THIS PUBLICATION MAY BE REPRODUCED OR
TRANSMITTED IN ANY FORM OR BY ANY MEANS, ELECTRONIC
OR MECHANICAL, INCLUDING PHOTOCOPY, RECORDING, OR ANY
INFORMATION STORAGE AND RETRIEVAL SYSTEM, WITHOUT
PERMISSION IN WRITING FROM THE PUBLISHER.

ACADEMIC PRESS, INC.
111 Fifth Avenue, New York, New York 10003

United Kingdom Edition published by
ACADEMIC PRESS, INC. (LONDON) LTD.
24/28 Oval Road, London NW1

Library of Congress Cataloging in Publication Data

Main entry under title:

Fundamentals of optical fiber communications.

    Bibliography: p.
    Includes index.
    1.   Fiber optics.   2.   Optical communications.
I.   Barnoski, Michael K.,   (date)
TA1800.F86   621.38'0414   75-44430
ISBN 0–12–079150–1

PRINTED IN THE UNITED STATES OF AMERICA

D
621.3804'14
FUN

Contents

5 PHOTODETECTORS FOR FIBER SYSTEMS

6 DESIGN OF REPEATERS FOR FIBER SYSTEMS

7 DESIGN CONSIDERATIONS FOR MULTITERMINAL NETWORKS

List of Contributors

Michael K. Barnoski, Hughes Research Laboratories, Malibu, California 90265

James E. Goell, ITT Electro-Optical Products Division, Roanoke, Virginia 24022

Donald Keck, Corning Glass Works, Corning, New York 14830

Henry Kressel, RCA Laboratories, Princeton, New Jersey 08540

S. D. Personick, Bell Telephone Laboratories, Holmdel, New Jersey 07733

Preface

The achievement of low loss transmission has made the optical fiber waveguide the leading contender as the transmission medium for a variety of future systems. This new low-loss medium offers the potential of many significant advantages compared with metallic conductors, including long distance transmission without repeaters, EMP, EMI, crosstalk and ground loop immunity, high bandwidth capabilities, small size and weight, high degree of intercept security and dielectric isolation, and long-term cost reduction. The foregoing desirable features of optical fiber waveguides have strongly simulated efforts in the supporting technologies such as fiber cabling, couplers, long-life solid state sources and high performance receivers. As a result of the considerable progress in these areas the utilization of optical fibers in military and commercial systems appears imminent.

The material presented in this book is intended to provide the reader with a tutorial treatment of fiber optic technology as applied to communications systems. Since this technology is in a rapid state of expansion and since much material is proprietary, a complete in-depth coverage of all aspects is difficult. As a result the emphasis in this text has been placed on the more fundamental considerations.

The editor gratefully acknowledges the contributing authors and the institutions with which they are associated for their wholehearted cooperation in the preparation of this book. Particular thanks are extended to the Hughes Research Laboratories, a division of the Hughes Aircraft Company.

The most competent assistance of Mrs. Barbara Schmaltz in the preparation of parts of the text is much appreciated. Grateful acknowledgement is also extended to the staff of the University of California, Santa Barbara, in particular, Mr. Larry Nicklin and Ms. Judy Weisman.

The editor is greatly indebted to Dr. James F. Lotspeich for his critical reading and useful comments on the content of the two chapters written by the editor.

M.K. Barnoski

Figure Credits

Page 42, Fig. C.3   From D. Cloge and E.J. Marcatili, *Bell Syst. Tech. J. 52* (1973), Fig. 6, p. 1571. Reproduced by permission.

Pages 52-54, Figs. D2-D4   From R. Olshansky, *Appl. Opt. 14* (1975), Figs. 1, 5, 6, pp. 941, 949, 944. Reproduced by permission.

Page 104, Figs. 3.12, 3.13   From F.L. Thiel, R.E. Love, and R.L. Smith, *Appl. Opt. 13* (1970), Figs. 1 and 2, p. 241. Reproduced by permission.

Page 111, Fig. 4.1   From C.J. Neuse, H. Kressel, and I. Landy, *IEEE Spectrum, 9* (1972), Fig. 1, p. 29. Reproduced by permission.

Page 126, Fig. 4.8   From R.D. Burnham, P.D. Dapkus, N. Holonyak, Jr., D.L. Keune, and H.R. Zwicker, *Solid State Electron, 13* (1970), Fig. 7, p. 204. Reproduced by permission.

Page 127, Fig. 4.9   From R.U. Martinelli and D.G. Fisher, *Proc. IEEE, 62* (1974), Fig. 14, p. 1349. Reproduced by permission.

Page 129, Fig. 4.11   From H. Kressel and I. Ladany, *RCA Review, 36* (1975), Fig. 4, p. 234. Reproduced by permission.

Page 138, Fig. 4.16   From C.A. Burrus and B.I. Miller, *Opt. Commun. 4* (1971), Fig. 1, p. 307. Reproduced by permission.

Page 139, Fig. 4.17   From H. Kressel and M. Ettenberg, *Proc. IEEE, 63* (1975), Fig. 1, p. 1360. Reproduced by permission.

Page 142, Fig. 4.19   From C.J. Neuse and G.H. Olsen, *Appl. Phys. Letters 26* (1975), Figs. 1 and 2, p. 529. Reproduced by permission.

Page 146, Fig. 4.21   From M. Ettenberg, H. Kressel, and H.F. Lockwood, *Appl. Phys. Letters 25* (1974), Fig. 2, p. 82. Reproduced by permission.

Page 208, Fig. 7.2   From M. DiDomenico, Jr., *Industrial Research* (August 1974), p. 50. Reproduced by permission.

Chapter 1 - OPTICAL FIBER WAVEGUIDES

D. B. Keck

Corning Glass Works

Corning, New York

Introduction

The achievement of low loss transmission coupled with other attendant advantages has made the optical fiber waveguide the leading contender as the transmission medium for a variety of systems ranging in length from 10 to 10,000 meters. Environmental aspects of optical fiber waveguides are receiving increased attention as systems come closer to reality and thus far the results are encouraging. While the primary interest today is in transmitting between opto-electronic devices on either end, the optical fiber waveguide, and its use in these systems, is a driving force for the entire field of integrated optics. The future of optical fiber waveguides is therefore extremely promising. In the ensuing pages the propagation characteristics of the optical fiber waveguide will be examined in some detail.

Many types of waveguiding structures have been proposed in the literature such as the single material and W-type guides, each with its own advantages and disadvantages. Throughout, however, the most universally applicable waveguide type is still the single high refractive index solid core surrounded by a lower refractive index solid cladding. It is this structure which will be examined, with particular emphasis on its long distance propagation characteristics.

This chapter is divided into four main sections. In the first section the ideal waveguide will be discussed. This will begin with a ray description to obtain an intuitive picture of the propagation, move on to the detailed

solution of Maxwell's equations for the cylindrical waveguide, and end with the WKB solution which furnishes a power-
ful approximate solution. The second section brings in a
degree of imperfection to the waveguide in the form of the
constituent materials and the various sources of waveguide
attenuation are discussed. Information carrying capacity
in the ideal waveguide is the topic of the third section
with the effect of the constituent waveguide materials being
included in the analysis. This discussion makes use of the
results of the WKB solution of the first section. The final
section deals with propagation in a perturbed waveguide
which leads to intermodal coupling. The perturbations may
result unintentionally from attempts to package the waveguide structure. The resulting mode coupling potentially
leads to an attenuation-bandwidth system tradeoff which is
discussed.

A - PROPAGATION IN IDEAL CYLINDRICAL WAVEGUIDES

Ray Theory

Since the early 1900's physics has been faced with the
duality of rays and waves. For most problems involving
electromagnetic propagation one has found that ray formalism,
while not incorrect, was not best suited for explaining the
details of the physical phenomenon involved. This is also
the case for the cylindrical fiber waveguide as we shall see.
Ray optics, however, does provide a simpler picture for
describing waveguide operation and therefore warrants discussion.

<u>Rays in step index fibers.</u> Let us consider first the
simple step index waveguide, which as shown in axial cross-
section in Fig. A.1, consists of a uniform core region of
diameter 2a and index n_1, surrounded by a cladding of index
n_2, where $n_1 > n_2$. Two types of rays exist in such a structure; meridional rays, which pass through the guide axis,
and skew rays, which do not. A typical meridional ray is
shown in the figure. As long as the external angle θ_0 that
it makes with the guide axis is less than θ_c where,

$$\sin \theta_c = (n_1{}^2 - n_2{}^2)^{\frac{1}{2}} \quad , \qquad\qquad (1A.1)$$

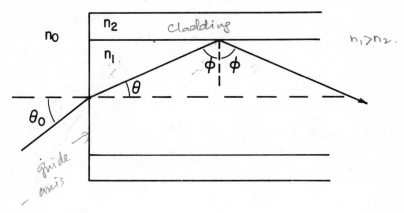

Fig. A.1   Typical meridional ray in a step refractive index
            fiber.

the ray will be <u>totally internally reflected at the core-</u>
cladding interface. In the traditional optical sense, the
numerical aperture of the waveguide is defined,

$$NA = \sin \theta_c \qquad \qquad (1A.2)$$

With simple geometry the path length of a meridional ray,
can also be obtained,

$$P(\theta) = L \sec \theta \qquad , \qquad\qquad (1A.3)$$

where L is the axial length of the guide. The path length
and therefore the transit time is a function of the angle of
the ray. This differential delay between the modes reduces
the information capacity of the waveguide. It also renders
the step index guide incapable of imaging.

The general skew ray is shown in Fig. A.2. It is
incident on the end of the guide at a point, $\vec{\rho}_0 = x\hat{i} + y\hat{j}$,
and at an angle defined by the unit vector $\vec{S}_0 = L_0\hat{i} + M_0\hat{j}$
$+ N_0\hat{k}$ where L_0, M_0, and N_0 are the direction cosines of the
ray. If $\vec{S}_m$ and $\vec{\rho}_m$ describe the ray prior to the mth reflection at the core-cladding interface, then one has the
vector relations,

$$(\vec{S}_m - \vec{S}_{m+1}) \times \vec{\rho}_m = \vec{0} \qquad , \qquad\qquad (1A.4)$$

3

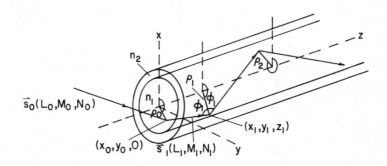

Fig. A.2   Typical skew ray in a step refractive index
            fiber.

$$(\vec{S}_m + \vec{S}_{m+1}) \cdot \vec{\rho}_m = 0 \quad . \tag{1A.5}$$

The first of these is the condition for coplanarity of the
incident and reflected ray while the second is the condition for equal angles of incidence and reflection. For
total internal reflection there is also the condition,

$$\vec{S}_m \cdot \frac{\vec{\rho}_m}{|\rho_m|} \leq \frac{n_1^{\,2} - n_2^{\,2}}{n_1^{\,2}} \quad . \tag{1A.6}$$

From these equations and with much algebra the general expression for the capture of a ray can be derived,

$$\left\{ L_0^{\,2} + M_0^{\,2} - \left(\frac{x_0 M_0 - y_0 L_0}{a} \right)^2 \right\}^{\frac{1}{2}} \leq NA \quad . \tag{1A.7}$$

If one considers a ray in the x-z plane, ($y_0 = 0$) then at
the point $x_0 = a$, Eq. (1A.6) is independent of M . That is
to say the ray could enter at $90°$ with respect to the y axis
and still be confined, obviously with zero axial velocity,
however. Skew rays can indeed exist at very high angles in
the waveguide. More will be said about these rays when
leaky modes are considered in a later section.

    Rays in graded index fibers. Attention is next turned
to the more general refractive index core and a discussion
of rays in such a medium. The general ray path described

4

by the vector $\vec{r}$ in Fig. A.3, it is found to be perpendicular to planes of constant $S(x,y,z)$ where

$$(\vec{\nabla S})^2 = n^2 \quad . \tag{1A.8}$$

This is the Eikonal equation which is derived from Maxwell's equations. If ds is an elemental path length along the ray then it can be shown that,

$$\frac{d}{ds}\left(n \frac{d\vec{r}}{ds}\right) = \vec{\nabla n} \quad . \tag{1A.9}$$

If the ray makes a small angle with respect to the z-axis the replacement, ds → dz, can often be made which corresponds to the paraxial approximation.

To match the waveguide geometry this equation is cast into cylindrical coordinates. The dependence of the unit vectors on the cylindrical coordinate variables must be taken into account during the differentiation. The following three component equations are obtained,

$$\frac{d}{ds}\left(n \frac{d\rho}{ds}\right) - n\rho\left(\frac{d\theta}{d\rho}\right)^2 = \frac{dn}{d\rho} \quad (\rho\text{-component}) \tag{1A.10a}$$

$$n\left(\frac{d\rho}{ds}\right)\left(\frac{d\phi}{ds}\right) + \frac{d}{ds}\left(n\rho \frac{d\phi}{ds}\right) = 0 \quad (\phi\text{-component}) \tag{1A.10b}$$

$$\frac{d}{ds}\left(n \frac{dz}{ds}\right) = 0 \quad (z\text{-component}) \cdot \tag{1A.10c}$$

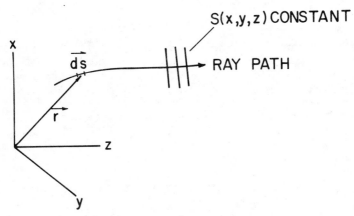

Fig. A.3   Coordinate system for describing a ray in an inhomogeneous medium.

5

It has been assumed in writing these that n is a function only of the radial coordinate. The z-component can be directly integrated to give

$$ds = (\frac{n}{n_0 N_0})\ dz \qquad . \qquad (1A.11)$$

Where as before N_0 is the direction cosine with respect to the z-axis, ρ_0 is the initial position of the ray and $n_0 \equiv n(\rho_0)$. Using this result the ϕ-component can be rewritten,

$$\rho^2 \frac{d\phi}{dz} = \frac{1}{N_0}\ (x_0 M_0 - y_0 L_0) \qquad . \qquad (1A.12)$$

Finally substituting both of these into the ρ-component and integrating once we obtain,

$$z = \int_{\rho_0}^{\rho} \frac{d\rho\ N_0}{\{\left[\frac{n(\rho)}{n_0}\right]^2 + (1 - \frac{\rho_0}{\rho})^2 (x_0 M_0 - y_0 L_0) - N_0^2\}^{\frac{1}{2}}} \qquad . \qquad (1A.13)$$

Thus the ray path can be uniquely specified once the index distribution, $n(\rho)$ and the initial ray parameters, x_0, y_0, L_0, and M_0 are known.

To obtain a feel for rays in a graded index medium we consider a few specific examples. Consider first the case of meridional rays. Without loss of generality we may pick $y_0 = M_0 = 0$ and then $x_0 = \rho_0$. Equation (1A.12) then becomes quite simply,

$$z = \int_{\rho_0}^{\rho} \frac{N_0 d\rho}{\{\left[\frac{n(\rho)}{n_0}\right]^2 - N_0^2\}^{\frac{1}{2}}} \qquad . \qquad (1A.14)$$

Consider first the distribution $n(\rho) = n_a(1 - (\alpha\rho)^2)^{\frac{1}{2}}$, the so called square law medium where $n_a \equiv n(0)$. Substituting this into Eq. (1A.14) and evaluating the elementary integral gives the radial coordinate of the ray as a function of position,

$$\rho = C \sin \frac{\alpha z}{n_0 N_0} \qquad , \qquad (1A.15)$$

where

$$C = \frac{n_0 N_0}{\alpha}\{\frac{1}{N_0^2 (1 + (\alpha\rho_0)^2)} - 1\}^{\frac{1}{2}} \qquad . \qquad (1A.16)$$

The ray path is periodic in z with a period,

$$\Lambda = \frac{2\pi N_0}{\alpha} \left(1-(\alpha\rho_0)^2\right)^{\frac{1}{2}}.$$ (1A.17)

Thus the period depends upon both the input position ρ_0 and input angle N_0 and is therefore different for every meridional ray. In general as long as the index exhibits a monotonic decrease with radius, a sinusoidal path within the waveguide will result with its period determined by the initial conditions and the exact nature of the profile.

Next consider the distribution $n(\rho) = n_a$ sech $(\alpha\rho)$. Again the integral in Eq. (1A.14) is of elementary form and the result is,

$$\sinh(\alpha\rho) = C' \sin(\alpha z) + \sinh(\alpha\rho_0) \quad ,$$ (1A.18)

where

$$C' = \left(\frac{\cosh^2(\alpha\rho_0)}{N_0} - 1\right)^{\frac{1}{2}} \quad .$$ (1A.19)

Once again the path is periodic in z. Now, however, the period is independent of the initial position or angle of the ray. This is the condition for focusing of the ray. The focal length is,

$$F = \frac{\pi}{2\alpha} \quad .$$ (1A.20)

This distribution has zero differential delay for all meridional rays and forms the basis for the "self-focusing" waveguides.

Thus far only meridional rays have been considered. There is one index distribution for which the path for a special class of skew rays can be obtained exactly (1). The rays are helical, with the condition that $d\rho/dz = 0$, and the index distribution is,

$$n(\rho) = n_a \left(1+(\alpha\rho)^2\right)^{-\frac{1}{2}} \quad .$$ (1A.21)

It is then found that the path of the ray is described by

$$\phi = \left[\frac{M_0}{N_0\rho_0}\right] z + \phi_0.$$ (1A.22)

7

Again for this ray class we see that the path is periodic in z with a length,

$$\Lambda = \frac{2_0 N_0 \rho_0}{M_0} \quad ,$$

(1A.23)

but that it again depends on the initial position and angle of the ray. Thus dispersion will again exist between various rays of this class.

In presenting these examples we have obtained a general picture of ray propagation in the general graded guide. It is to be noted that no single distribution is capable of simultaneously focussing all rays and that therefore differential delays will always exist. It should also be noted that if α is small, these three index distribution functions can be expanded. To first order in α, they are the same,

$$n(\rho) = n_a \left(1-(\alpha\rho)^2\right)^{\frac{1}{2}} \quad .$$

(1A.24)

This indicates that with small α a near focus condition can be obtained for all rays resulting in small differential delays and therefore high information carrying capacity.

In the section on information carrying capacity we will see that a slight modification of this profile indeed results in low dispersion.

Mode Theory

More detailed knowledge of propagation characteristics of the optical fiber waveguide can only be obtained by solution of Maxwell's equations. This leads to only certain allowed modes which can propagate in a particular dielectric structure. If the structure is such that a large number of modes can propagate, this theory can become very complex and intractable. One will search for simplifications and approximations to the exact theory.

Since many other publications have discussed mode theory of various structures in great detail, our approach will be merely to outline the general method and the results. We will then consider two simplifying techniques which give more insight into electromagnetic propagation in dielectric cylinders.

One begins as always with Maxwell's equations. Assuming a linear, isotropic material in the absence of currents and charges, they become,

$$\vec{\nabla} \times \vec{E} = \frac{\partial \vec{B}}{\partial t} \qquad \vec{\nabla} \times \vec{H} = \frac{\partial \vec{D}}{\partial t} \qquad\qquad (1A.25)$$

$$\vec{\nabla} \cdot \vec{B} = 0 \qquad\quad \vec{\nabla} \cdot \vec{D} = 0 \qquad\qquad ,$$

with the constitutive relations,

$$\vec{D} = \varepsilon \vec{E} \qquad\qquad \vec{B} = \mu \vec{H} \qquad\qquad . \qquad\qquad (1A.26)$$

By taking the curl of the first two equations and applying a vector identity, these can be reduced to the scalar wave equation,

$$\nabla^2 \Psi = \varepsilon \mu \frac{\partial^2 \Psi}{\partial t^2} \qquad\qquad , \qquad\qquad (1A.27)$$

where Ψ represents each component of $\vec{E}$ and $\vec{H}$. In doing this the assumption must be made that $\nabla \varepsilon / \varepsilon = 0$. Marcuse (2) shows that if the change in this term is small over a distance of one wavelength this term may be neglected. This is the case for the problems to be considered.

A cylindrical coordinate system ρ, ϕ, and z is defined with the z axis coaxial with the waveguide. Transforming the curl equations to cylindrical coordinates results in two sets of three equations for the components of $\vec{E}$ and $\vec{H}$ in terms of one another. One can solve these for the transverse components E_ρ, E_ϕ, H_ρ, and H_ϕ in terms of E_z and H_z. We seek solutions which are harmonic in time and z,

$$\begin{pmatrix} \vec{E} \\ \vec{H} \end{pmatrix} = \begin{pmatrix} \vec{E}(\rho,\phi) \\ \vec{H}(\rho,\phi) \end{pmatrix} e^{-i(\omega t - \beta z)} \qquad\qquad , \qquad\qquad (1A.28)$$

where β is the z-component of the propagation vector. With this, the equations for the transverse field components can be written,

$$E_\rho = \frac{-i}{K^2} \left(\beta \frac{\partial E_z}{\partial \rho} + \frac{\mu \omega}{\rho} \frac{\partial H_z}{\partial \phi} \right) \qquad\qquad (1A.29a)$$

$$E_\phi = \frac{-i}{K^2} \left(\frac{\beta}{\rho} \frac{\partial E_z}{\partial \rho} - \mu \omega \frac{\partial H_z}{\partial \rho} \right) \qquad\qquad (1A.29b)$$

9

$$H_\rho = \frac{-i}{K^2} \left(\beta \frac{\partial H_z}{\partial \rho} - \frac{\mu\omega}{\rho} \frac{\partial E_z}{\partial \phi} \right) \tag{1A.29c}$$

$$H_\phi = \frac{-i}{K^2} \left(\frac{\beta}{\rho} \frac{\partial H_z}{\partial \rho} + \omega\varepsilon \frac{\partial E_z}{\partial \rho} \right) \quad , \tag{1A.29d}$$

where

$$K^2 = k^2 - \beta^2 = \left(\frac{2\pi n}{\lambda}\right)^2 - \beta^2 \quad . \tag{1A.30}$$

Here k is the propagation constant in a medium of dielectric constant ε or alternatively refractive index n.

The scalar wave equation must now be solved for E_z and H_z to complete the solution. Equation (1A.27) is expressed in cylindrical coordinates and the variables are separated by assuming

$$\begin{pmatrix} E_z \\ H_z \end{pmatrix} = A \, F(\rho) \, e^{i\nu\phi} \quad . \tag{1A.31}$$

From the differential equations for ϕ it is found that ν must be an integer in order to ensure azimuthal periodicity. The differential equation for $F(\rho)$ becomes

$$\frac{\partial^2 F}{\partial \rho^2} + \frac{1}{\rho} \frac{\partial F}{\partial \rho} + \left(k^2 - \beta^2 - \frac{\nu^2}{\rho^2}\right) F = 0 \quad . \tag{1A.32}$$

This equation must be solved for β and $F(\rho)$ subject to the boundary conditions of a specific waveguide structure.

<u>Solution for the step index waveguide.</u> One of the few refractive index distributions for which Eq. (1A.32) can be solved is that of a homogeneous core of index n_1 and radius a, surrounded by an infinite cladding of index n_2. Then the solutions are Bessel functions appropriately chosen to assure finite $F(\rho)$ at $\rho = 0$ and $F(\rho) \to 0$ as $\rho \to \infty$.

For $\rho < a$, this is a J-type Bessel function of order ν so that,

$$\begin{pmatrix} E_z \\ H_z \end{pmatrix} = \begin{pmatrix} A \\ B \end{pmatrix} J_\nu(u\rho) \, e^{i\nu\phi} \quad , \tag{1A.33}$$

where $u^2 = (k_1^2 - \beta^2)$, $k_1 = 2\pi n /\lambda$, and A and B are arbitrary constants. For the region $\rho > a$ one must use a modified Hankel function,

$$\begin{bmatrix} E_z \\ H_z \end{bmatrix} = \begin{bmatrix} C \\ D \end{bmatrix} K_\nu(w\rho) \, e^{i\nu\phi} \quad , \tag{1A.34}$$

where $w^2 = \beta^2 - k_2{}^2$, $k_2 = 2\pi n_2/\lambda$, and C and D are again constants.

The quantity,

$$V^2 = (u^2 + w^2)a^2 = \left(\frac{2\pi a}{\lambda}\right)^2 (n_1{}^2 - n_2{}^2) \quad , \tag{1A.35}$$

is constant of the waveguide and gives much information concerning its operation.

Several points can be determined from Eqs. (1A.33) and (1A.34). As $w\rho \to \infty$, $K_\nu(w\rho) \to e^{-w\rho}$. For the proper behavior as $\rho \to \infty$, $w > 0$. Thus we see that $\beta \geq k_2$. The equality represents the cutoff condition at which point the propagation is no longer bound to the core region. Inside the core u must be real, and therefore $k_1 \geq \beta$. So we find the allowed range for the propagation constant for bound solutions is,

$$k_2 \leq \beta \leq k_1. \tag{1A.36}$$

The exact solution for β must come from satisfying the boundary condition that the tangential components of E and H be continuous at the boundary $\rho = a$. This condition gives four homogeneous equations in the unknown constants A, B, C and D. Only if the determinant of the coefficients vanishes will a solution exist. After much algebra this results in the eigenvalue equation for β,

$$\left(\frac{J'_\nu(ua)}{uJ_\nu(ua)} + \frac{K'_\nu(wa)}{wK_\nu(wa)}\right)\left(\frac{k_1{}^2 J'_\nu(ua)}{uJ_\nu(ua)} + \frac{k_2{}^2 K'_\nu(wa)}{wK_\nu(wa)}\right) \tag{1A.37}$$

$$= \nu^2\beta^2 \left(\frac{1}{u^2} + \frac{1}{w^2}\right) \quad .$$

The primes indicate differentiation with respect to the argument. When this equation is solved for β, only discrete values within the range allowed in Eq. (1A.36) will be found.

Consider first the case $\nu = 0$. In this case the fields of the dielectric cylinder break into TM($H_z = 0$) and TE($E_z = 0$) modes just as in the case of the conducting cylinder. Since

11

$\nu = 0$, the modes are radially symmetric. Because of the oscillatory behavior of $J_\nu(u\rho)$ there will be m roots of this equation for a given ν, subject to the constraint, $\beta_{\nu m} \geq k_2$. These modes correspond to a uniform density of meridional rays, making the same discrete angle with respect to the z-axis.

For $\nu \neq 0$ the situation is more complex. Then hybrid modes, designated $HE_{\nu m}$ and $EH_{\nu m}$, exist for which both E_z and H_z are non-zero. As before there exist m roots for a given ν value. The designation HE or EH is given depending on whether H_z and E_z makes the larger contribution to the transverse field.

An important mode parameter is its cutoff frequency. The following equations give the cutoff conditions for the various mode types,

$$\left. \begin{array}{c} EH_{\nu m} \\ HE_{1m} \end{array} \right\} J_\nu(u_m a) = 0 \tag{1A.38a}$$

$$HE_{\nu m} \ (n_1^2+1)J_{\nu-1}(u_m a) = \frac{u_m a}{(\nu-1)} J_\nu(u_m a) \nu = 2,3,4\ldots\ldots \tag{1A.38b}$$

$$\left. \begin{array}{c} TE_{0m} \\ TM_{0m} \end{array} \right\} J_0(u_m a) = 0 \ . \tag{1A.38c}$$

There is one mode, designated HE_{11}, for which no cutoff exists. This is the basis for the single-mode waveguide. By adjusting the guide parameters such that the next higher modes, TE_{01}, TM_{01}, HE_{21}, are cutoff, only the HE_{11} is left to propagate. This occurs for,

$$2.405 > \frac{2\pi a}{\lambda} (n_1^2-n_2^2) = V \ . \tag{1A.39}$$

A plot of the normalized propagation constant β/k, for a few of the low order modes is shown in Fig. A.4.

Let us next look at the field distribution of modes. For one polarization, the electric field component in the z direction is,

$$E_z \sim J_\nu(u_m\rho) \cos \nu\phi, \tag{1A.40}$$

whereas the transverse components are obtained from Eq. (1A.29),

12

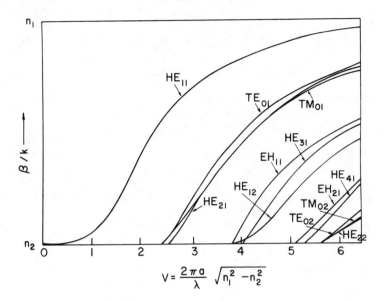

Fig. A.4.   Normalized propagation constant as a function of V-parameter for a few of the lowest order modes of a step waveguide.

$$E_\rho \sim \pm J_{\nu\pm1}\left(u_m\rho\right)\cos\nu\phi \qquad (1A.41)$$

$$E_\phi \sim J_{\nu\pm1}\left(u_m\rho\right)\sin\nu\phi \quad , \qquad (1A.42)$$

where the + and − correspond to the $EH_{\nu m}$ and $HE_{\nu m}$ modes respectively. The transverse field is then given by,

$$\vec{E}_t = E_\rho\hat{\rho} + E_\phi\hat{\phi}$$

$$\sim J_{\nu\pm1}\left(u_m\rho\right)\left(\pm\cos\nu\phi\ \hat{\rho} + \sin\nu\phi\ \hat{\phi}\right) \quad . \qquad (1A.43)$$

From this the field patterns of various modes as well as admixtures of them can be generated. Since,

$$-\hat{1} = -\cos\phi\ \hat{\rho} + \sin\phi\ \hat{\phi} \quad , \qquad (1A.44)$$

13

it is seen for example that the HE_{1m} modes are linearly polarized. Also the TE_{0m}, TM_{0m} are seen to be independent of angle, and therefore radially symmetric. In general the field pattern obtained will be a complex admixture of the fields of the various modes. The HE_{11} mode, however, is found to be simply proportional to $J_0(u\rho)$.

There are no azimuthal zeros in the fields except through linear combinations of modes. Schematically, the first two sets of modes appear as in Fig. A.5.

<u>Weakly guiding solutions.</u> The equations just examined are exact solutions for the homogeneous core waveguide. Snyder (3) and Gloge (4) have recognized the fact that because $n_1 \sim n_2$ there is a great similarity between the eigenvalues for the $HE_{\nu+1,m}$ and the $EH_{\nu-1,m}$ modes which in fact are equal if $n_1 = n_2$. This suggests a possible linear combination of solutions to achieve a simplification.

The assumption is made that for the region $\rho < a$,

HE$_{11}$

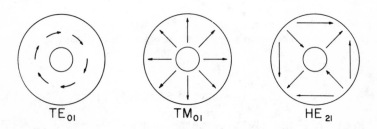

TE$_{01}$ TM$_{01}$ HE$_{21}$

Fig. A.5   Schematic diagram of the electric field vector
            for the four lowest order modes of a step wave-
            guide.

$$E_z = \frac{iAu}{2} \left\{ J_{\nu+1}(u\rho) \begin{bmatrix} \sin (\nu+1)\phi \\ -\cos (\nu+1)\phi \end{bmatrix} + J_{\nu-1}(u\rho) \begin{bmatrix} \sin(\nu-1) \\ -\cos(\nu-1) \end{bmatrix} \right\}$$

$$H_z = \frac{iAu}{2k} \left(\frac{\varepsilon}{\mu}\right)^{\frac{1}{2}} \left\{ J_{\nu+1}(u\rho) \begin{bmatrix} \cos (\nu+1)\phi \\ \sin(\nu+1)\phi \end{bmatrix} - J_{\nu-1}(u\rho) \begin{bmatrix} \cos(\nu-1)\phi \\ \sin(\nu-1)\phi \end{bmatrix} \right\} .$$

$$(1A.45)$$

As before $u^2 = k_1^2 - \beta^2$ and $k_1 = 2\pi n_1/\lambda$. The transverse fields E_ρ, E_ϕ, H_ρ, and H_ϕ are the quantities desired. These may be obtained from Eq. (1A.29). It is more instructive to cast them into cartesian coordinates which may be accomplished by the transformation matrix,

$$\begin{bmatrix} E_x \\ E_y \end{bmatrix} = \begin{bmatrix} \cos \phi & -\sin \phi \\ \sin \phi & \cos \phi \end{bmatrix} \begin{bmatrix} E_\rho \\ E_\phi \end{bmatrix} .$$

$$(1A.46)$$

A similar matrix equation exists for the magnetic field. Upon laboriously evaluating the x and y components of E and H and making the assumption that $n \approx n_1 \approx n_2$ it will be found for the assumed E_z and H_z, that $E_x = H_y = 0$ and that,

$$E_y = A\, J_\nu(u\rho) \begin{bmatrix} \cos \nu\phi \\ \sin \nu\phi \end{bmatrix}$$

$$H_x = -n\, A\left(\frac{\varepsilon}{\mu}\right)^{\frac{1}{2}} J_\nu(u\rho) \begin{bmatrix} \cos \nu\phi \\ \sin \nu\phi \end{bmatrix} .$$

$$(1A.47)$$

Thus it is seen that this linear combination represents a wave linearly polarized in the y direction. This form of the transverse field is obviously simpler than the exact solutions in Eq. (1A.43).

For a complete description of the field one of course requires the mode with orthogonal polarization. This is obtained by making the replacement $\sin(\nu\pm1)\phi \to \cos(\nu\pm1)\phi$, $-\cos(\nu\pm1)\phi \to \sin(\nu\pm1)\phi$ and $J_{\nu-1} \to -J_{\nu-1}$ for E_z and $\cos(\nu\pm1)\phi \to \sin(\nu\pm1)\phi$, $\sin(\nu\pm1)\phi \to -\cos(\nu\pm1)\phi$ and $-J_{\nu-1} \to J_{\nu-1}$ for H_z in Eq. (1A.45). Then it will be found that; $E_y = H_x = 0$ and

$$E_x = A\, J_\nu(u\rho) \begin{bmatrix} \cos \nu\phi \\ \sin \nu\phi \end{bmatrix}$$

$$H_y = nA\left(\frac{\varepsilon}{\mu}\right)^{\frac{1}{2}} J_\nu(u\rho) \begin{bmatrix} \cos \nu\phi \\ \sin \nu\phi \end{bmatrix} .$$

$$(1A.48)$$

15

Similar linear combinations are assumed for E_z and H_z in the region $\rho > a$ except the Bessel functions are replaced by a modified Hankel function of imaginary argument, $K_\nu(w\rho)$. Upon doing this again it is found that only linearly polarized fields exist. These have the form,

$$E_y = \frac{A\ J_\nu(ua)}{K_\nu(wa)}\ K_\nu(w\rho) \begin{Bmatrix} \cos \nu\phi \\ \sin \nu\phi \end{Bmatrix}$$

$$H_x = -n\ A\left[\frac{\varepsilon}{\mu}\right]^{\frac{1}{2}} \frac{J_\nu(ua)}{K_\nu(wa)}\ K_\nu(w\rho) \begin{Bmatrix} \cos \nu\phi \\ \sin \nu\phi \end{Bmatrix} \qquad (1A.49)$$

$$E_x = H_y = 0 \qquad ,$$

for the E_z and H_z in Eq. (1A.45) and,

$$E_x = \frac{A\ J_\nu(ua)}{K_\nu(ua)}\ K_\nu(w\rho) \begin{Bmatrix} \cos \nu\phi \\ \sin \nu\phi \end{Bmatrix}$$

$$H_y = \frac{n\ A\ J_\nu(ua)}{K_\nu(wa)}\ K_\nu(w\rho) \begin{Bmatrix} \cos \nu\phi \\ \sin \nu\phi \end{Bmatrix} \qquad (1A.50)$$

$$E_y = H_x = 0 \qquad ,$$

for the orthogonal polarization. While these expressions give the form of the field in the various regions of space, we must yet satisfy the boundary condition that the tangential components E_ϕ, E_z, H_ϕ and H_z be continuous across the boundary $\rho = a$. The fields given in expressions (1A.47), (1A.48), (1A.49) and (1A.50) already are such that the components are continuous at least to the degree that $n_1 \sim n_2$. The continuity of the z component requires equating like components of $\sin(\nu\pm1)\phi$ and $\cos(\nu\pm1)\phi$. From this we get the eigenvalue equation which must be satisfied for a solution to exist,

$$\frac{u\ J_{\nu\pm1}(ua)}{J_\nu(ua)} = \mp \frac{w\ K_{\nu\pm1}(wa)}{K_\nu(wa)} \qquad . \qquad (1A.51)$$

This equation represents a considerable simplification from that given in Eq. (1A.37). Snyder (3) has shown it to be accurate to within 1% and 10% for $\Delta \leq 0.1$ and $\Delta \leq 0.25$ respectively. We must remember, however, that each solution,

$\beta_{\nu m}$, of this equation is really twofold degenerate compared with the exact solution because of the assumption made concerning the fields and that $n_1 \approx n_2$. We must also keep in mind that these field configurations do not really exist at all points along an actual waveguide since there is always a slight dispersion between the $HE_{\nu \pm 1, m}$ and $EH_{\nu \pm 1, m}$ modes from which they are derived. This causes these modes to combine alternately in and out of phase with one another thus never continually exhibiting the above field distributions.

The notation for labelling these linearly polarized modes obviously is no longer the same as for the exact solution since the integer ν now refers to the combination of exact modes with labels $\nu+1$ and $\nu-1$. The lowest order mode (HE_{11}) now has the propagation constant labelled β_{01}.

The coefficient A in the expression for the fields is obtained by evaluating the Poynting vector for a mode. This is simply the total power in the mode passing through infinite cross section and thus serves to normalize the mode field.

The assumptions made above become less valid near cutoff and since one desires simplified solutions for the multimode case, it is reasonable to look at approximations far from cutoffs. It has been shown (5) in that case,

$$u_{\nu m} = u_m^{\infty}(1 - \frac{2\nu}{V})^{\frac{1}{2}\nu} \quad , \tag{1A.52}$$

where u_m^{∞} is the mth root of the equation

$$J_{\nu}(u_m^{\infty}a) = 0 \quad , \tag{1A.53}$$

and $V = 2\pi a/\lambda (n_1^2 - n_2^2)^{\frac{1}{2}}$ is the characteristic guide parameter. For the case of the HE_{1m} mode ($\nu = 0$), the limit of Eq. (1A.52) quite simply becomes,

$$u_{0m} = u_m^{\infty} \exp(-1/V) \quad . \tag{1A.54}$$

The value of the modal propagation constant, $\beta_{\nu m}$, can readily be obtained from Eq. (1A.54).

It is instructive to estimate the total number of modes for a given value of V. This is obtained by counting

17

the total number of roots of Eq. (1A.53) for a given value of ν subject to the condition that

$$u_m^\infty a \leq V \quad .$$

(1A.55)

It is known that for large m the roots are approximately given by,

$$u_m^\infty a \approx (\nu + 2m)\pi/2 \leq V \quad .$$

(1A.56)

For $m = 0$, $\nu = 2V/\pi$ and for $\nu = 0$, $m = V/\pi$. We see that these two points define a triangle in the ν-m plane. Each point within the triangle represents 4 degenerate modes. The total number of modes is therefore given by 4X the area of the triangle times the density of modes, $N = 4V^2/\pi^2 \approx 0.4V^2$. A better approximation has been shown to be (4),

$$N = \frac{V^2}{2} \quad .$$

(1A.57)

Finally we look at the normalized total power for a given mode in the core and cladding regions. This is obtained by integrating the Poynting vector over each region. Gloge (4) shows that these are given quite simply in this field description by,

$$\frac{P_{core}}{P_T} = 1 - \frac{u^2}{V^2} \left\{ 1 + \frac{J_\nu^2(ua)}{J_{\nu+1}(ua)J_{\nu-1}(ua)} \right\}$$

$$\frac{P_{clad}}{P_T} = 1 - \frac{P_{core}}{P_T} \quad .$$

(1A.58)

Far from cutoff Marcuse (5) has obtained the expression for the power in the cladding,

$$\frac{P_{clad}}{P_T} = \left(\frac{u_m^\infty a}{V}\right)^4 \left(1 - \frac{2}{V}\right) \quad .$$

(1A.59)

This clearly shows that as V increases, the fraction of power carried in the cladding for any mode decreases. So, for example, for the HE_{11} mode one can calculate that for $V = 1$ approximately 70% of the power resides in the cladding while at $V = 2.405$, where the next mode group begins, the situation is reversed and about 84% of the power travels within the core.

The field distribution in the cladding behaves as $K_\nu(w\rho)$ as seen in Eqs. (1A.49 - 1A.50). For large ρ asymptotic form is $K_\nu(w\rho) \to \exp(w\rho)$. Then when $\rho = 1/w$ the field will have decayed to $1/e$ of its maximum value. Defining this as the mode radius, $\rho_{\nu m}$ and applying Eq. (1A.52) gives,

$$\rho_{\nu m} = \frac{1}{w} = \frac{a}{\{V^2 - (u_m^\infty a)^2 (1 - \frac{2\nu}{V})^{\frac{1}{2}}\}^{\frac{1}{2}}} \qquad . \qquad (1A.60)$$

For the HE_{11} mode for example, with $V = 1$, the mode radius is $\rho_{01} \approx 3a$. One must therefore have a cladding on the guide in excess of this to avoid perturbation of the field.

Thus far only the homogeneous core waveguide surrounded by an infinite cladding has been considered. There is one other index distribution of significance to optical fibers for which an exact solution to the scalar wave equation can be obtained. That is the square law medium whose refractive index has the form,

$$n^2(\rho) = n^2(0)\left(1 - 2\Delta(\frac{\rho}{a})^2\right) \qquad , \qquad (1A.61)$$

where $n(0)$ is the axial index of refraction, Δ is approximately the fractional index difference between core and cladding and a is the core radius. The solution to this problem has been given (2) and will not be presented in detail here. The field solutions for this problem are the well-known Laguerre-Gauss functions. The propagation constant for the modes in this case is simply given,

$$\beta_{pq} = n(0)k\{1 - \frac{2\sqrt{2}\Delta}{n(0)ka} (p+q+1)\}^{\frac{1}{2}} \qquad . \qquad (1A.62)$$

It is seen that if $m = p+q$, the modes are m-fold degenerate.

WKBJ solution for graded index. One would like to find solutions to the dielectric waveguide problem which are both simpler and also applicable to more general refractive index distributions. This can be done by using WKBJ method which is well-known from quantum mechanics (6). The method has been applied to the solution of the dielectric waveguide problem by Kurtz and Streifer (7) and more recently Gloge and Marcatili (8). It is a particularly useful technique, especially for obtaining the propagation constant, and bears a close examination. It generally ignores anomalies of modes near cutoff.

Recall that if the index distribution is a function only of ρ, that the scalar wave Eq. (1A.27) is separable in cylindrical coordinates and has the formal solution given in Eq. (1A.31). The differential equation for the radial component of the wave equation must be solved,

$$\frac{d^2\Psi}{d\rho^2} + \frac{1}{\rho}\frac{d\Psi}{d\rho} + \left(k^2(\rho) - \beta^2 - \frac{\nu^2}{\rho^2}\right)\Psi = 0 \quad , \tag{1A.63}$$

where $\Psi(\rho)$ represents either E_z or H_z. The radial wave-number is,

$$k(\rho) = \frac{2\pi}{\lambda}n(\rho) \quad , \tag{1A.64}$$

where $n(\rho)$ is the radial index distribution. As before ν is the azimuthal mode number and β is the axial component of the propagation vector. The general approach with the WKBJ method is to recognize that if n = constant, the general solution would be a superposition of plane waves,

$$\Psi(\rho) \sim e^{iU(\rho)} \quad . \tag{1A.65}$$

This solution, is substituted into Eq. (1A.63) yielding,

$$i\frac{d^2U}{d\rho^2} - \left(\frac{dU}{d\rho}\right)^2 + \frac{i}{\rho}\frac{dU}{d\rho} + \left(k^2(\rho) - \beta^2 - \frac{\nu^2}{\rho^2}\right) = 0. \tag{1A.66}$$

If the index variation with ρ is slow so that the function $U(\rho)$ is nearly constant over a distance of one wavelength, it can be expanded in a power series in $k^{-1} = (2\pi/\lambda)^{-1}$,

$$U(\rho) = U_0(\rho) + \frac{1}{k}U_1(\rho) + \dots \quad . \tag{1A.67}$$

Substituting this into Eq. (1A.66) and gathering like powers of k to first order gives the following equations,

$$-\left(\frac{dU_0(\rho)}{d\rho}\right)^2 + \left\{k^2(\rho) - \beta^2 - \frac{\nu^2}{\rho^2}\right\} = 0 \tag{1A.68}$$

$$ik\left(\frac{d^2U_0}{d\rho^2}\right) - 2\left(\frac{dU_0}{d\rho}\right)\left(\frac{dU_1}{d\rho}\right) + i\frac{k}{\rho}\left(\frac{dU_0}{d\rho}\right) = 0 \quad . \tag{1A.69}$$

Since we shall be primarily concerned with the propagation constant, only the zero order approximation is needed. For a field description $U_1(\rho)$ would also be required. Integrating Eq. (1A.68) gives,

$$U_0(\rho) = \int \{k^2(\rho) - \beta^2 - \frac{\nu^2}{\rho^2}\}^{\frac{1}{2}} d\rho \quad . \tag{1A.70}$$

Using the quantum mechanical analog, the quantity $k^2(\rho)$ $- \nu^2/\rho^2$ represents a potential well within which a particle of energy β is constrained to move. The potential term, $-\nu^2/\rho^2$, may be thought of as coming from a centrifugal force and represents the energy associated with the angular motion of the particle. Only if U_0 is real will one have an oscillating solution for $\Psi(\rho)$ and therefore a bound mode. This requires that the radical in Eq. (1A.70) be positive. In general there exist two values ρ_1 and ρ_2 for which the radical vanishes. Outside these two values or turning points, U_0 becomes imaginary leading to decaying fields.

It can be shown for bound solutions that the phase U_0 (evaluated between the turning points) must be approximately a multiple of π,

$$m\pi \approx \int_{\rho_1}^{\rho_2} \{k^2(\rho) - \beta^2 - \frac{\nu^2}{\rho^2}\}^{\frac{1}{2}} d\rho \quad , \tag{1A.71}$$

where $m = 0,1....$ is the radial mode number which counts the number of half periods between the turning points.

Shown schematically in Fig. A.6 is a plot of the radical in Eq. (1A.70) or the potential function for the propagating plane waves. Bound modes will be found for any $\beta \geq k(a)$. For a given value of β the meaning of the two turning points is obvious. It is between these two radii that the ray associated with the assumed plane wave solution is constrained to move. For a given β, as ν increases, a point will be reached such that the turning points merge and beyond which the wave is no longer bound. We may count the number of bound modes which can exist above a given value of β by summing Eq. (1A.71) over all ν values,

$$m(\beta) = \frac{4}{\pi} \int_{\rho_1}^{\rho_2} \int_0^{\nu_{MAX}} \{k^2(\rho) - \beta^2 - \frac{\nu^2}{\rho^2}\} d\rho d\nu \quad . \tag{1A.72}$$

Here it has been assumed that ν is large enough so that the summation can be replaced by integration. The factor of 4 comes from the degeneracy of the modes with respect to

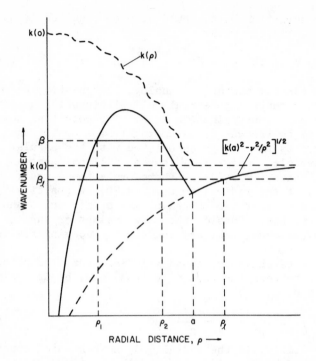

Fig. A.6. Schematic plot of the waveguide potential as a function of radial distance. The lines $k(\rho)$ and $k(a)$ represent the core and cladding wave numbers respectively. β and β_ℓ represent propagation constants for a bound and a leaky mode respectively.

polarization and orientation. In order to count all the modes, the lower turning must go to $\rho = 0$. This will occur for $\nu = 0$. Upon integrating with respect to ν one obtains,

$$m(\beta) = \int_0^{\rho_2} \{k^2(\rho) - \beta^2\}\rho d\rho \qquad . \qquad (1A.73)$$

To proceed further with this analysis, specific information regarding the index profile is needed. A particularly useful form is (8),

$$n^2(\rho) = n^2(0)\{1-2\Delta(\rho/a)^\alpha\} \qquad , \qquad (1A.74)$$

where $\Delta = n^2(0)-n^2(a)/2n^2(0)$, a is the core radius and α specifies the shape of the profile. For $\alpha = \infty$ we have simply the step profile, while for $\alpha = 2$ the square law profile results.

The turning point ρ_2 occurs when $k(\rho_2) = \beta$. Thus from Eq. (1A.74) we obtain

$$\rho_2 = a\{\frac{1}{2\Delta}\left[1-\left(\frac{\beta}{kn(0)}\right)^2\right]\}^{\frac{1}{2}} \qquad . \qquad (1A.75)$$

Upon integration of Eq. (1A.73) with this upper limit, the number of modes with propagation constants greater than becomes,

$$m(\beta) = \left[\frac{k^2n^2(0)-\beta^2}{2\Delta k^2n^2(0)}\right]^{\frac{2+\alpha}{\alpha}} (\frac{\alpha}{\alpha+2})a^2k^2n^2(0)\Delta \qquad . \qquad (1A.76)$$

The quantity $m(\beta)$ is almost the mode number since it simply counts the modes up to β. We must recall, however, that to this approximation the modes will fall into degenerate groups. Recalling the case $\alpha = 2$, if p is the integer specifying the mode group, each group is p-fold degenerate. Summing the degeneracy of the mode group from zero to the pth level gives the total number of modes above that level, $m(\beta)$. Thus one has,

$$\sum_0^p p \approx p^2 = m(\beta) \qquad . \qquad (1A.77)$$

All bound modes lie above $\beta = n(a)k$ and thus the total number of modes from Eq. (1A.76) is,

$$M = (\frac{\alpha}{\alpha+2})a^2k^2n^2(0)\Delta \qquad . \qquad (1A.78)$$

The total number of mode groups is simply $P = \sqrt{M}$. Using this expression with Eq. (1A.77) yields the desired expression for the axial propagation constant as a function of mode number,

$$\beta_p = kn(0)\{1-2\Delta(\frac{p}{P})^{\frac{2\alpha}{\alpha+2}}\}^{\frac{1}{2}} \qquad . \qquad (1A.79)$$

This expression is extremely useful and will be used extensively to discuss dispersion and mode coupling in the waveguide. Let us use the preceding results to obtain a few features of the graded index waveguide.

The total number of modes is simply the phase space volume and therefore measures the light gathering capability of the waveguide. It is seen from Eq. (1A.78) that the total number of modes in the parabolic waveguide, $\alpha = 2$, is 1/2 of the number of the step waveguide, $\alpha = \infty$. Thus the parabolic waveguide accepts only 1/2 the light of the step waveguide.

Another interesting quantity which will be needed for mode coupling work involves the mode group spacing, $\delta\beta$. This is given by taking the derivative,

$$\frac{d\beta_p}{dp} = \delta\beta = \left(\frac{\alpha}{\alpha+2}\right)^{\frac{1}{2}} \frac{2\sqrt{\Delta}}{a} \left(\frac{p}{P}\right)^{\frac{\alpha-2}{\alpha+2}} . \tag{1A.80}$$

It is seen that the mode spacing for the step waveguide increases linearly with increasing mode number. On the other hand, for the parabolic waveguide, $\alpha = 2$, the spacing is independent of the mode number as was found earlier.

If θ is the angle between the mode wave vector and the waveguide axis, then,

$$\beta_p = kn(\rho)\cos\theta_p . \tag{1A.81}$$

The magnitude of β depends on the radial coordinate. An easy visualization shows that the maximum angle which the wavevector makes with the axis occurs for $\rho = 0$ at which point,

$$\beta_\rho = kn(0)\left(1-\sin^2\theta_p\right)^{\frac{1}{2}} = kn(0)\left(1-2\Delta\left(\frac{p}{P}\right)^{\frac{2\alpha}{\alpha+2}}\right)^{\frac{1}{2}} . \tag{1A.82}$$

This gives the relation

$$\sin\theta_p = \sqrt{2\Delta}\left(\frac{p}{P}\right)^{\frac{\alpha}{\alpha+2}} . \tag{1A.83}$$

24

For the case of the step guide,

$$\sin\theta_p = \sqrt{2\Delta}\,\frac{p}{P} \quad , \qquad\qquad 1A.84)$$

and we see there is a perfect correspondence between the
mode angle and mode number. When $p = P$, $\sin\theta_p = \sqrt{2\Delta} =$
$\sin\theta_c$ (Eq. (1A.1)). The far field pattern from a step
waveguide therefore corresponds exactly to the mode spectrum of the waveguide. For any other value of α however,
Eq. (1A.83) shows that the mode angle and mode number are
not linearly related. This merely reflects the fact that
in a graded index guide no single angle uniquely characterizes a mode.

A considerable amount of information concerning a
waveguide can be deduced from the near and far field
intensity distributions. First consider the near field.
The limiting value of β_p is $kn(a)$. Thus from Eq. (1A.81)
the maximum angle θ_c a wavevector can possess at a point
ρ is,

$$\cos\theta_c(\rho) = \frac{n(a)}{n(\rho)} \quad . \qquad\qquad (1A.85)$$

A local numerical aperture can be defined,

$$NA(\rho) = n(\rho)\,\sin\theta_c(\rho) = \left(n^2(\rho) - n^2(a)\right)^{\frac{1}{2}} \quad . \qquad (1A.86)$$

Assuming equal excitation of all modes, the power accepted at the point ρ relative to $\rho = 0$ will be,

$$\frac{p(\rho)}{p(0)} = \frac{NA^2(\rho)}{NA^2(0)} = \frac{n^2(\rho) - n^2(a)}{n^2(0) - n^2(a)} \quad . \qquad (1A.87)$$

For the index distribution in Eq. (1A.74) one obtains,

$$\frac{p(\rho)}{p(0)} = 1 - (\frac{\rho}{a})^{\alpha} \quad . \qquad\qquad (1A.88)$$

If no losses occur during transmission this quantity is
simply the near field output from the fiber. Thus the
refractive index profile parameter can potentially be
measured by launching a Lambertian distribution into a
short length of waveguide and measuring the emergent
near field power distribution.

25

The far field is obtained by assuming that all points on the end face emit light uniformly into a cone defined by $\theta_c(\rho)$. The total power received at θ from all points on the end of the waveguide is simply,

$$\frac{P(\theta)}{P(0)} = \{1 - \frac{\sin^2\theta}{2\Delta n^2(0)}\}^{2/\alpha} \quad .$$

(1A.89)

Once α is known from the near field, the far field pattern gives information concerning $2\Delta n^2(0) = \sin^2\theta_c(0)$ which is the maximum angle a wavevector can have with respect to the axis and still be accepted by the waveguide.

B - ATTENUATION IN FIBER WAVEGUIDES

The level of understanding and reduction of attenuation in fiber waveguides is such that it is now possible to consider transmission links on the order of 10 km in length over a fairly wide spectral region. The sources of loss are broadly grouped into categories, absorption and radiation, and can originate from either the material or waveguide structure.

In general, these two types of loss will be different for the core and cladding region. We have already seen in Eq. (1A.58) that the power carried in the core and cladding for the step guide is a function of mode number. Thus if one assumes that the loss coefficients are α_1 and α_2 in the core and cladding respectively, the total loss for the vm^{th} mode will be,

$$\alpha_{vm} = \frac{\alpha_1 \, P_{vm}^{core} + \alpha_2 \, P_{vm}^{clad}}{P_{vm}^{total}} \quad .$$

(1B.1)

The total loss of the waveguide will obviously be obtained by summing over all modes weighted by the fractional power in that mode. The situation for a graded profile guide is far more complex. Then both the loss coefficient and the modal power can be functions of the radial coordinate so that one must integrate over all ρ to obtain the modal loss prior to summing over the modes,

$$\alpha_{\nu m} = \frac{\int_0^\infty \alpha_{\nu m}(r) \, P_{\nu m}(r) r dr}{P_{\nu m}^{total}} \qquad (1B.2)$$

Owing to the complexity the multimode waveguide, no correlation of this model with experiment has been obtained. It has been observed however, that the loss is different for various modes as indicated by the model. The loss is in general found to increase with increasing mode number (9). In this section the loss effects of mode coupling will be neglected and a discussion of them postponed to a later section. Here we shall concern ourselves with identifying the other contributing factors to loss in waveguides and will indicate their magnitude.

Absorption loss. Absorption loss in glasses can come from three factors, intrinsic absorption of the basic material, impurity absorption, and atomic defect absorption. Intrinsic absorption originates due to charge transfer bands in the ultraviolet region and vibration or multiphonon bands in the near infrared. If these bands are sufficiently strong their tails will extend into the spectral region of interest for fiber communications, 700-1100 nm. For most glasses considered for optical fibers, the vibration bands in the near infrared are both far enough removed, occuring in the 8-12 μm wavelength region, and not sufficiently strong to cause a problem. The ultraviolet bands are far stronger and potentially more troublesome. For the case of a germanium doped silica, however, Urbach's rule has been applied to the band edge (10) and it was shown that for wavelengths greater than about 600 nm, less than 1 dB/km absorption resulted.

Metal ions in glass are traditional sources of impurity absorption. Initially these were most feared and many studies on bulk glasses showed that the allowed levels for such things as Fe, Cu, V and Cr could not exceed 8, 9, 18 and 8 ppb respectively in order to obtain sub-20 dB/km loss at band center. High silica waveguides are regularly made, however, such as the one whose spectrum is shown in Fig. B.1 in which these impurities do not contribute to the loss. Waveguides made from more conventional glasses may still suffer absorption due to these, however, the impurity levels

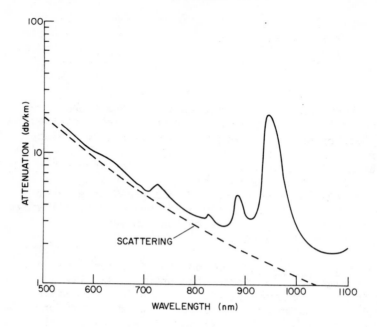

Fig. B.1   Attenuation spectrum for a high silica wave-guide between 500 and 1100 nm. The dotted curve indicates the total scattering. Clearly visible are the OH absorption bands.

required are so small that no direct correlation with wave-guide absorption is possible. The only impurity for which a direct correlation has been shown is the OH radical, whose bands at 725, 825, 875 and 950 nm are clearly visible in the figure. These are overtones and combination bands of the fundamental OH vibration at 2.73 μm and of silica matrix. The strength of the 950 nm band has been shown to be approx-imately 1 dB/km/ppm. All absorption in the spectrum shown can be accounted for by OH absorption.

Atomic defect absorption is induced by a stimulus such as the thermal history or by intense radiation of the glass. The magnitude of such induced losses can be quite large as, for example, in the case of titanium doped silica where a reduction, $Ti^{4+} \rightarrow Ti^{3+}$, occurs during fiberization to produce losses of several thousand dB/km (11). Similarly, radiation induced losses of ~ 20,000 dB/km are possible for conventional fiber optic glasses from gamma radiation levels of 3000 rads. Generally, however, one can choose glasses which are less susceptible to these effects. For example, germanium doped silica has been shown to exhibit an attenuation at 820 nm of only ~ 16 dB/km for a radiation level of 4300 rads (12). Such things as background cosmic radiation for example would have a small long term effect on such a waveguide.

Scattering loss. All transparent materials scatter due to frozen in thermal fluctuations of constituent atoms. These cause density and hence index variations within the material. This intrinsic scattering is believed to represent the fundamental limit to attenuation in waveguides. It can be calculated by subdividing the sample into small volumes which act as dipoles. One then sums all the dipoles integrated over all angles and can relate the scattering loss to the isothermal compressibility β,

$$\alpha_s = \frac{8\pi^3}{3\lambda^4} (n^2 - 1)kT\beta \qquad , \qquad (1B.3)$$

where T is the transition temperature at which the fluctuations are frozen into the glass. This loss decreases very rapidly with increasing wavelength. For fused silica using a transition temperature of $1500^\circ C$ one calculates a loss of 1.7 dB/km at 820 nm which is in good agreement with experiment. A tradeoff between the transition temperature and the compressibility for a given material has been shown to exist (13). In fact Li-Al-SiO$_2$ glasses have been measured to have losses due to density fluctuations several times less than that of fused silica.

For waveguides one generally has glasses with more than one oxide in which case another form of scattering occurs. This is due to concentration fluctuations in the constituent oxides and also causes a loss. The expression for this loss has been given,

$$\alpha = \frac{16\pi^3 n}{3\lambda^4} \left(\frac{dn}{dc}\right)^2 \overline{(\Delta c)^2} \, \delta V \qquad , \qquad (1B.4)$$

in which $(\Delta c)^2$ is the mean square concentration fluctuation and δV is the volume over which it occurs. Generally the magnitude of the index fluctuation is not known and therefore the scattering cannot be calculated. Rather the scattering is used to obtain the index fluctuation. It is generally observed that if the added oxide raises the refractive index, larger fluctuations tend to occur. Thus for high index glasses, losses due to concentration fluctuations tend to dominate those due to density fluctuations. However, for high silica glasses, concentration fluctuations typically account for only ~ 25% of the observed scattering loss.

In addition to these two intrinsic scattering loss mechanisms one can induce scattering through non-linear effects, such as stimulated Raman and Brillouin scattering. Because of the small core size, the confined guidance and the long interaction length, relatively low absolute power levels are required to observe such effects (14). For long lengths of fiber, expressions for the maximum power at 1.06 μm for these two procceses have been derived (15),

$$P_{MAX} \sim 1 \times 10^{12} \, A\alpha \quad \text{(watts)} \, \text{(Backward Raman)}$$

$$P_{MAX} \sim 5 \times 10^9 \, A\alpha \quad \text{(watts)} \, \text{(Backward Brillouin)},$$

(1B.5)

where $A(cm^2)$ is the power carrying area and $\alpha(cm^{-1})$ is the loss coefficient. These expressions predict for example, that threshold powers of 500 and 2.5 watts would be required for Raman and Brillouin scattering respectively for a 75 μm core diameter fiber with a 4 dB/km linear attenuation. Due to the rapid dissipation of energy by these two processes, at these thresholds the fiber loss would show a rapid increase. The above equations have assumed the pump bandwidth equal to the scattering process bandwidth. This will generally not be the case, $\Delta\nu(\text{Brillouin}) < \Delta\nu(\text{Raman}) < \Delta\nu \, (\text{laser})$, and hence considerably higher thresholds will occur in practice. In actual fact 2000 watts has been injected into a 2 km length of guide having a 75 μm core diameter with no non-linear attenuation observed (16).

In addition to the above scattering loss mechanisms one can typically have radiation losses associated with the waveguide structure. We have assumed the cladding to be infinitely thick while in practice it is in the order of a few tens of microns. Thus if the jacket is lossy, as for example, to minimize crosstalk, some fraction of the mode fields can reach this region and be attenuated. If n_3 is the index of the jacket region which occurs at a distance $\rho = \rho_j$, then the rate of outward power flow to axial power flow is simply,

$$\eta = \left\{ \frac{(kn_3)^2 - \beta^2}{\beta^2} \right\}^{\frac{1}{2}} \quad . \tag{1B.6}$$

Multiplying this by the relative power at the cladding-jacket interface gives the loss coefficient for the νm^{th} mode,

$$\alpha_{\nu m}(dB) = 4.34 \; \pi \rho_j \; \frac{P_{\nu m}(\rho_j)\eta}{P_{\nu m}^{total}} \quad , \tag{1B.7}$$

where $p(\rho_j)$ is the power density at the interface and $P_{\nu m}^{total}$ is the total power in the mode. To obtain the total loss one must obviously sum over all modes. This is a very complex computation. Gloge (17) has obtained an approximate expression for the cladding thickness necessary to restrict the loss of a fraction, F, of modes to less than 1 dB/km. He finds,

$$\rho_j \approx a\left(1 + \frac{36}{\sqrt{FV^3}}\right) \quad . \tag{1B.8}$$

For a typical multimode guide with $V = 50$, a jacket radius of $\rho_j = 1.4a$ would be required for 90% of the modes to have less loss than 1 dB/km.

Thus far only bound modes have been considered in our discussions. The restriction for these modes, $\beta \geq n_2 k$, strictly holds only for $\nu = 0$. For $\nu \neq 0$ however, a mode with $\beta < n_2 k$ only turns purely radiative beyond the radius,

$$\rho_\ell = \nu\left((kn_2)^2 - \beta_\ell^2\right)^{-\frac{1}{2}} \quad . \tag{1B.9}$$

For large ν, the leakage of the evanescent field through the region $a < \rho < \rho_\ell$ can be small, and hence these modes can propagate long distances. Pask, et.al. (18) have derived an expression for the loss coefficient for these modes for large ν, in the step index fiber,

$$\alpha_\ell = \frac{4}{\pi a \sqrt{2\Delta}} \; \frac{\theta_z^2}{V} \; \frac{1}{|K_\nu(wa)|^2} \quad , \tag{1B.10}$$

where $\theta_z = \cos^{-1}(\beta/n_1 k)$. Upon evaluating this expression a number of modes will be found with a loss coefficient, $\alpha_\ell < \alpha_0$. An approximate expression for this number is (17),

$$N = (.05)V^2 \left[\frac{a\alpha_0}{\sqrt{\Delta}}\right]^{1/V} . \qquad (1B.11)$$

For a typical multimode guide with $\Delta = 0.01$ and $a = 35$ μm, for $\alpha_0 = 1$ dB/km this would represent a fractional increase in mode volume of approximately 7%. The total number of leaky modes is given by (18),

$$N = \frac{V^2}{2} \{1 - \frac{8}{3\pi} \sqrt{2\Delta} + ...\} . \qquad (1B.12)$$

For small Δ the number of leaky modes represents a significant fraction of the number of guided modes and hence can have a significant effect on propagation.

An additional form of loss is present when the axis of the guide is curved. This occurs because, for a mode to maintain an equiphase front on a radial plane, the energy beyond a critical radius, R_c, would have to exceed the phase velocity of light in the medium (19). The fraction of light beyond R_c is therefore radiated away. This effect has been observed and shown to be in qualitative agreement with theory for single mode waveguides (20). For multimode waveguides the situation is far more complex and to date no good correlation between theory and experiment have been obtained. The general observation is that for bend radii in excess of ~ 10 cm negligible loss occurs (21).

C - PULSE BROADENING IN OPTICAL FIBER WAVEGUIDES

In this section the effects of waveguide dispersion will be considered. This will be done primarily from the standpoint of digital transmission in which case pulse broadening produced by the waveguide is of primary concern. Later sections of this book will formally relate the amount of pulse broadening to the information carrying capacity of a waveguide transmission system.

The approach will be to first obtain a general expression for pulse broadening, and to apply the results of the WKB section to evaluate the result. It will be found that for a certain radial index profile pulse broadening can be

made very small giving rise to large information capacity.
However, a fundamental restriction placed by material dispersion will always limit the information capacity. The
complication of mode coupling between the modes will be
ignored here and postponed until the next section.

Intermodal and Intramodal Broadening

It has been seen that the modes of a waveguide are
specified by two integers ν and m which enumerate the azimuthal and radial modes of the electromagnetic fields of
the mode. The propagation constant $\beta_{\nu m}$ is found to depend
on the axial index n, the fractional index difference Δ
and the core radius a, and the propagating wavelength λ.
For considering pulse propagation one is concerned with the
group velocity or alternatively the group delay per unit
length for a given mode,

$$T_{\nu m} = \frac{1}{v_g} = \frac{1}{c} \frac{d\beta_{\nu m}}{dk} \qquad (1C.1)$$

where $k = 2\pi/\lambda$ is the free space constant. If the waveguide is excited by an impulse, its response at the point z
for a given wavelength λ will be obtained by a summation
over all modes,

$$P(t,z,\lambda) = \Sigma P_{\nu m}(\lambda,z)\delta\{t-z\,T_{\nu m}(\lambda)\} \quad . \qquad (1C.2)$$

The function $P_{\nu m}(\lambda,z)$ describes the power as a function of
wavelength and z for the given mode. In general this will
be a function of the specific form of the exitation at z = 0
as well as the attenuation characteristics of the mode as a
function of z. As stated earlier, the effects of power
coupling between modes will temporarily be neglected. Most
detectors are incapable of resolving individual wavelengths
so the quantity of practical interest is,

$$P(t,z) = \int_0^\infty d\lambda P(t,z,\lambda) \qquad (1C.3)$$

A brief digression is required at this point. A parameter is needed to specify the amount of pulse associated
with the pulse distribution P(t,z). For systems analysis,

the fourier transform provides perhaps the most useful tool,

$$\tilde{P}(\omega,z) = \int_{-\infty}^{\infty} P(t,z)e^{-i\omega t}dt \quad . \tag{1C.4}$$

Alternatively, however, the moments $M_n(z)$ of the distribution $P(t,z)$ defined by the equation,

$$M_n = \int_0^{\infty} t^n P(t,z)dt \quad , \tag{1C.5}$$

can do the same task. Often knowledge of only the first few moments provides sufficient information and hence this specification is considerably simpler. From Eqs. (1C.4) and (1C.5) one can show that,

$$M_n = (i\omega)^n \frac{d^n}{d\omega^n} \tilde{P}(\omega,z) \quad . \tag{1C.6}$$

Hence the complete set of moments is completely identical to specifying the fourier transform.

Applying the moment definition to Eq. (1C.2) and integrating over all wavelengths yields,

$$M_n(z) = z^n \int_0^{\infty} \Sigma P_{\nu m}(\lambda,z)T_{\nu m}^n(\lambda)d\lambda \quad . \tag{1C.7}$$

The main wavelength dependence in the mode power will come from the source spectral power distribution $S(\lambda)$ which is generally sharply peaked. It therefore can reasonably be assumed that,

$$P_{\nu m}(\lambda,z) \simeq S(\lambda)P_{\nu m}(z) \quad . \tag{1C.8}$$

Using the equation defining the moments, the following definitions can be made,

$$S_0 = \int_0^{\infty} S(\lambda)d\lambda$$

$$\lambda_0 = \frac{1}{S_0} \int_0^{\infty} S(\lambda)d\lambda \tag{1C.9}$$

$$\sigma_s = \frac{1}{S_0} \{ \int_0^{\infty} (\lambda-\lambda_0)^2 S(\lambda)d\lambda \}^{\frac{1}{2}} \quad ,$$

where S_0 is the total source power, λ_0 is the mean operating wavelength and σ_s is the rms spectral source width.

If the group index is a smoothly varying function in the vicinity of λ_0, then the group delay $T_{\nu m}$ may be expanded into a Taylor series,

$$T_{\nu m}(\lambda) = T_{\nu m}(\lambda_0) + T'(\lambda-\lambda_0) + T''(\lambda-\lambda_0)^2 + \ldots \quad (1C.10)$$

The primes denote the derivative with respect to λ evaluated at λ_0. Substituting this into Eq. (1C.7) and utilizing the definitions (1C.9) gives an expression for the nth moment,

$$M_n(z) \simeq z^n \Sigma P_{\nu m}(z) \left(T_{\nu m}(\lambda_0) + \frac{\sigma_s}{2\lambda_0^2} \{nT_{\nu m}^{n-1}(\lambda_0)\lambda_0^2 T''_{\nu m} \right.$$

$$\left. + n(n+1)T^{n-2}(\lambda_0)\lambda_0^2(T')^2\} \right) \quad . \quad (1C.11)$$

Since $\sigma_s/\lambda_0 \ll 1$, terms of order $(\sigma_s/\lambda_0)^3$ and higher may be neglected. This expression may be more simply written if we define $<A>$ as the average of the variable A over the mode distribution,

$$<A> \equiv \Sigma P_{\nu m}(z)A_{\nu m}/M_0 \quad . \quad (1C.12)$$

Then a normalized moment m_n can be defined,

$$m_n \equiv \frac{M_n}{M_0} = z^n \left(<T^n(\lambda_0)> + \frac{\sigma_s^2}{2\lambda_0^2} \{n\lambda_0^2<T^{n-1}T''> \right.$$

$$\left. + n(n+1)\lambda_0^2<T^{n-2}(T')^2>\} \right) \quad . \quad (1C.13)$$

With this definition, $m_0 = 1$. The average arrival time $T(z)$ is simply given,

$$T(z) = m_1 = z\left(<T(\lambda_0)> + \frac{\sigma_s}{2\lambda_0^2}<\lambda_0^2 T''> \right) \quad . \quad (1C.14)$$

The quantity in square brackets is the reciprocal group velocity for the guide structure.

For the purpose of specifying the information carrying capacity, the rms pulse width $\sigma(z)$ is most useful (22). With the above definitions it becomes,

35

$$\sigma^2(z) = m_2 - m_1^2 = \sigma^2_{\text{intermodal}} + \sigma^2_{\text{intramodal}}, \quad (1C.15)$$

where,

$$\sigma^2_{\text{intermodal}} \equiv z^2 \{ <T^2(\lambda_0)> - <T(\lambda_0)>^2$$

$$+ \left(\frac{\sigma_s}{\lambda_0}\right)^2 \left(<\lambda_0^2 T''(\lambda_0)T(\lambda_0)> - <\lambda_0^2 T''(\lambda_0)><T(\lambda_0)> \right) \}, \quad (1C.16)$$

and

$$\sigma^2_{\text{intramodal}} \equiv z^2 \left(\frac{\sigma_s}{\lambda_0}\right)^2 <\lambda_0^2 T'(\lambda_0)^2> \quad . \quad (1C.17)$$

The rms width has been separated into two components for the purpose of clarifying the physical situation. Both are dependent on the length of the waveguide. The intermodal broadening Eq. (1C.16) results from differences in the group delay between the various modes. The portion of this term which depends upon the relative source spectral width is small and can usually be neglected. The intramodal term, Eq. (1C.17) represents the average pulse broadening within each mode. This term provides the only non-vanishing contribution for a single mode guide and thus defines the ultimate limitation on information capacity. It is found to contain two distinct parts, one arising from the bulk material, the other from the guide structure. The separation can be made explicit by writing the mode delay time,

$$T_{\nu m} = N/c + \delta T_{\nu m} \quad , \quad (1C.18)$$

where N is the group index, N/c represents the delay common to all modes and $\delta T_{\nu m}$ is the correction introduced by the guide structure. Then, taking the derivative of Eq. (1C.18) with respect to wavelength gives,

$$T'_{\nu m} = -\lambda n'' + \delta T'_{\nu m} \quad , \quad (1C.19)$$

where n is the guide refractive index. Inserting this into Eq. (1C.17) yields,

$$\sigma^2_{\text{intramodal}} \equiv z \left(\frac{\sigma_s}{\lambda_0}\right)^2 \{ \lambda_0^2 n'' - 2\lambda_0 n'' <\lambda_0 \delta T'>$$

$$+ <(\lambda_0 \delta T')> \} \quad . \quad (1C.20)$$

The first term is a pure material effect, the last a pure waveguide effect and the middle a mixed material waveguide term. It is found that the latter two terms are negligible and hence for most practical wavelengths the intramodal term is simply,

$$\sigma_{intramodal} = z\left(\frac{\sigma_s}{\lambda_0}\right) \lambda_0^2 n'' \qquad . \qquad (1C.21)$$

This term defines the ultimate information capacity limit for all types of waveguides. It can be reduced by decreasing the relative source spectral width. Alternatively for a given glass composition, it may be possible to operate at wavelengths for which n'' is reduced.

Evaluation of Pulse Broadening in Grade Index Waveguides

To carry the analysis further, expressions for the group delay must be obtained. This can most simply be done using the results of the WKB solution obtained in the previous section. For the pth mode group the propagation constant β_p is given in Eq. (1A.79). Considering explicit and implicit k dependence the group delay is given from Eq. (1C.1),

$$T_p = N\{1+\Delta(\frac{\alpha-2-\varepsilon}{\alpha+2})(\frac{p}{P})^{\frac{2\alpha}{\alpha+2}} +$$

$$\Delta^2 \frac{(3\alpha-2-2\varepsilon)}{2(\alpha+2)} (\frac{p}{P})^{\frac{4\alpha}{\alpha+2}}\} + 0\Delta^3 \qquad , \qquad (1C.22)$$

where

$$\varepsilon = -\frac{2n}{N} \frac{\lambda}{\Delta} \frac{d\Delta}{d\lambda} \qquad . \qquad (1C.23)$$

The quantity ε takes into account the varying dispersion properties between the core and cladding materials.

From this it is seen that if $\alpha > 2 + \varepsilon$, higher order mode groups will have a greater relative delay. The opposite is true of $\alpha < 2 + \varepsilon$. To first order in Δ, the group delay difference between modes vanishes if $\alpha = 2 + \varepsilon$. Since ε is generally small this indicates that the near parabolic index profile tends to minimize the intermodal dispersion. It is recalled that in discussing ray propagation the parabolic profile resulted in a near focussing

37

condition. It is thus seen that the perfect focus and minimum dispersion condition are synonomous.

Approximating the summation in Eq. (1C.12) with an integration and assuming equal exitation of the modes, the intermodal and intramodal contributions to the rms width are simply,

$$\sigma_{intermodal} = \frac{LN\Delta}{2c} \left(\frac{\alpha}{\alpha+1}\right)\left(\frac{\alpha+2}{3\alpha+2}\right)^{\frac{1}{2}} \{C_1{}^2 + \frac{4C_1C_2\Delta(\alpha+1)}{2\alpha+1}$$

$$+ \frac{4\Delta^2 C_2{}^2}{(5\alpha+2)}\frac{(2\alpha+2)^2}{(3\alpha+2)} \}^{\frac{1}{2}} \quad , \tag{1C.24}$$

and,

$$\sigma_{intramodal} = \frac{L\sigma_s}{c\lambda} \{(\lambda n'')^2 - 2(\lambda n'')^2 N\Delta C_1\frac{\alpha}{\alpha+1}$$

$$+ (N\Delta)^2 C_1{}^2 \left[\frac{2}{3}\frac{\alpha}{\alpha+2}\right] \}^{\frac{1}{2}} \quad , \tag{1C.25}$$

where

$$C_1 = \frac{\alpha-2-\varepsilon}{\alpha+2}$$

$$C_2 = \frac{3\alpha-2-2\varepsilon}{2(\alpha+2)} \quad . \tag{1C.26}$$

The minimum intermodal delay is found to occur for

$$\alpha_c = 2 + \varepsilon - \Delta \frac{(4+\varepsilon)(3+\varepsilon)}{5 + 2\varepsilon} \quad . \tag{1C.27}$$

The correction to α_c involving Δ comes about because a partial cancellation of the mode dependent term in Eq. (1C.24) can be found. The correction produced by ε has been evaluated for the case of a TiO_2 doped silica core - silica clad waveguide (23). It is found that this term can cause 10% - 15% differences from the perfect parabolic, $\alpha = 2$, profile.

For the case of the optimal profile and considering light sources in the 700-1100 nm wavelength region it is found that the intramodal dispersion term will always

dominate. Fig. C.1 shows a plot of the total rms pulse
width ss a function of the parameter α for three different
GaAs sources. These correspond to an LED, an injection
laser and a distributed feedback laser having rms spectral
widths of 150 Å, 10 Å and 2 Å respectively. The dotted
curve shows the results if material dispersion were neglected. The effect of diminishing the source spectral
width is readily apparent. The information capacity limit
for these sources will be in the range 0.13, 2 and 10 Gbits-
km/sec respectively.

Since the quantity ε is a function of wavelength, the
value α_c will also be a function of wavelength. Alternatively
for a given α profile, the rms pulse broadening will be a
function of wavelength. This behavior is shown in Fig. C.2
having been calculated for the waveguide above in the range

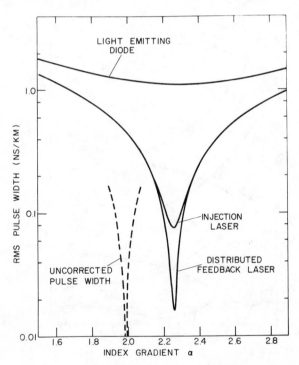

Fig. C.1   RMS pulse width as a function of the index
gradient parameter α for three possible light sources.

1.7<α<2.9. Thus it will be possible to tune the source
wavelength slightly to optimize guide performance.

A comment should be made concerning the dependence
of the information capacity on the relative core-cladding
index difference Δ. As shown in Eq. (1C.24) the intermodal broadening increases directly as Δ for $\alpha \neq \alpha_c$. In
the vicinity of $\alpha = \alpha_c$, the broadening becomes proportional
to Δ^2. Thus a tradeoff will exist between increasing Δ to
obtain better source coupling and decreasing Δ to reduce
intermodal dispersion.

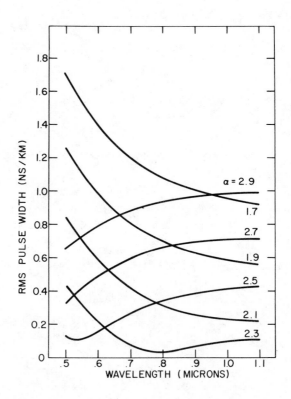

Fig. C.2 RMS pulse width as a function of wavelength for
several values of the index gradient parameter α.

Pulse Shape. At this point only an expression for the
width of a pulse has been obtained. For system calculations
the exact shape of the pulse is also of importance. In general, this will depend strongly upon differential mode
attenuation as well as the specific excitation conditions
of the waveguide. However, if uniform modal excitation is
assumed and no differential attenuation exists, then a
rather simple expression for the pulse shape can be obtained. This gives a reasonable qualitative description
and hence warrants examination. By inverting Eq. (1C.22)
and retaining only terms of order Δ, one obtains an expression for the mode group number in terms of its relative
group delay δT,

$$p = P\{\left[\frac{\alpha+2}{\alpha-2-\epsilon}\right] \frac{1}{\Delta}\}^{\frac{\alpha+2}{2\alpha}} (\delta T)^{\frac{\alpha+2}{2\alpha}} .$$ (1C.28)

The relative amount of power δI arriving per unit time is
simply,

$$\delta I = \left[\frac{2p}{p^2} \frac{dp}{d(\delta T)}\right] = \{\left[\frac{\alpha+2}{\alpha-2-\epsilon}\right] \frac{1}{\Delta}\}^{\frac{\alpha+2}{\alpha}} \left(\frac{\alpha+2}{\alpha}\right)(\delta T)^{2/\alpha} . $$ (1C.29)

The above expression holds only for $\alpha \neq 2 + \epsilon$. As this
condition is violated, higher order terms in Δ must be
retained. This equation is plotted in Fig. C.3. For $\alpha = \infty$,
the arriving power is independent of the relative delay
resulting in a gate function. The width of the gate, obtained as the difference between the earliest and latest
arrival time from Eq. (1C.22) is exactly as would be calculated from a meridional ray analysis, $\Delta t = LN\Delta/c$. For
$\alpha \neq \infty$, since the higher order modes carry a larger fraction
of the energy, the pulse is peaked at the relative delay
of the highest order mode. Thus for $\alpha > 2 + \epsilon$ this occurs
later than the zero delay. The opposite is true for $\alpha < 2 + \epsilon$.

Lastly, the relative arrival time as a function of mode
angle is considered. Using Eq. (1A.83), the group delay in
Eq. (1C.22) can be written,

$$T_p = \frac{LN}{c} \{1+\left[\frac{\alpha-2-\epsilon}{\alpha+2}\right] \frac{\theta^2}{2} + \left(\frac{3\alpha-2-2\epsilon}{\alpha+2}\right) \frac{\theta^4}{8} + ...\} .$$ (1C.30)

As already stated, the correspondence between mode number
and angle is strictly only good for the step guide in which
case this equation becomes,

41

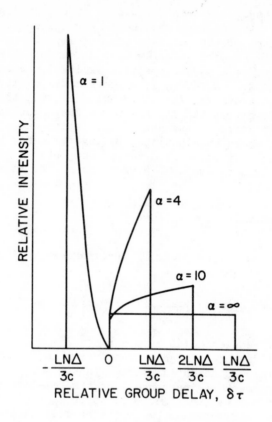

Fig. C.3. Relative pulse shape as a function of the index gradient parameter α. (After Gloge and Marcatili (8)).

$$T_p = \frac{LN}{c} \left\{ 1 + \frac{\theta^2}{2} + \frac{3\theta^4}{8} + .. \right\} = \frac{LN}{c} \sec \theta \quad . \qquad (1C.31)$$

It is seen that this is the same as Eq. (1A.3) if the latter is multiplied by the group velocity of the guide. Thus for the step waveguide, a meridional ray analysis gives a correct picture for intermodal pulse broadening.

D - MODE COUPLING IN FIBER WAVEGUIDES

Having thus far considered propagation aspects in the unperturbed waveguide we now turn to consider the effect which perturbations can have on waveguide performance. These perturbations can take the form of deviations of the fiber axis from straightness, fluctuations in the fiber diameter or refractive index variations, all as a function of distance along the waveguide. The perturbations may be intrinsic to the guide or produced by an external stress such as might be encountered in packaging of the waveguide. In any case the effect is to couple energy from one mode to another in a fashion prescribed by the specific perturbation. There is both benefit as well as loss associated with this effect. The benefit arises because the propagation delays associated with the modes are averaged, thus reducing the intermodal dispersion. On the other hand, energy is also coupled between bound and unbound modes manifesting itself in an additional form of attenuation. As will be shown, these two effects result in a loss-bandwidth product which, for a given perturbation, is constant and may be used to characterize the mode coupling.

Power Flow in Multimode Waveguides

Quite generally Maxwell's equations must be solved for the mode fields subject to the specific distortion of the waveguide. This involves solutions of an infinite set of coupled differential equations which are usually very difficult to solve. The solution also contains more information than is often desired. A considerable simplification can be obtained by considering only the coupled power flow (24) in the guide. Consider a guide having M modes. The total change in power $P_m(z,t)$ of the mth mode, as a function of distance will be,

$$\frac{dP_m}{dz} = \frac{\partial P_m}{\partial z} + \frac{\partial t}{\partial z}\frac{\partial P_m}{\partial t} = \frac{\partial P_m}{\partial z} + \frac{1}{v_m}\frac{\partial P_m}{\partial t} \quad , \qquad (1D.1)$$

where v_m is the group velocity of the mode. This change in power is due to three factors:

a) the power dissipated from the mode through normal loss and scattering,

43

b)   the power lost to other modes through coupling,

c)   the power gained from other modes through coupling.

The coupled power flow equation may then be written,

$$\frac{\partial P_m}{\partial z} + \frac{1}{v_m} \frac{\partial P_m}{\partial t} = -\gamma_m P_m + \sum_{n=1}^{M} d_{mn}(P_m - P_n) \quad , \qquad (1D.2)$$

where d_{mn} is the coupling coefficient from the mode m to n and is assumed reciprocal, and γ_m is the normal loss coefficient of the mode. This represents M coupled differential equations which in general have no analytic solution. To obtain a simplification, coupling is assumed to be dominated by transitions for which $n = m \pm 1$. This adjacent mode coupling is quite reasonable for most practical cases. Additionally the modes are assumed to be very closely spaced so that a transformation to a single continuous variable may be made. If the average coupling coefficient between the p and p+1 mode group is $d(p)$, then Eq. (1D.1) may be rewritten,

$$\frac{\partial P(p)}{\partial z} + \frac{1}{v(p)} \frac{\partial P(p)}{\partial t} = -\gamma(p)P(p) + \frac{1}{p} \frac{\partial}{\partial p} pd(p) \frac{\partial P(p)}{\partial p} \quad .$$

$$(1D.3)$$

This equation is seen to be a diffusion equation for the power in the pth mode group. Its solution can proceed once the functions $v(p)$, $\gamma(p)$ and $d(p)$ are specified. For the class of index profiles in Eq. (1A.74) the group velocity for the pth mode group can be obtained from Eq. (1C.2). A reasonable assumption for $\gamma(p)$ based on experimental evidence is that,

$$\gamma(p) = \gamma_0 \qquad\qquad\qquad p \le P_c$$
$$\gamma(p) = \infty \qquad\qquad\qquad P_c \le P \le P(\alpha) \qquad , \qquad (1D.4)$$

where P_c allows for the fact that only some fraction of the total number of mode groups $P(\alpha)$, Eq. (1A.78) might propagate. These restrictions give one boundary condition which will be needed for solution of Eq. (1D.3),

$$P(p) = 0 \qquad p > P_c \qquad . \qquad (1D.5)$$

The other boundary condition is obtained from the fact that the "diffusion current" at $p = 0$ is zero.

$$I(0) \equiv pd(p) \left.\frac{\partial P(p)}{\partial p}\right|_{p=0} = 0 \qquad . \qquad (1D.6)$$

This of course represents the fact that no power can flow to modes with $p < 0$.

Derivation of the coupling coefficient is very complex (25) and here we merely outline the main parts. Assume the refractive index distortion can be expressed by the fairly general function,

$$\delta n^2(\rho,\phi,z) = g(\rho)\cos(n\phi+\psi)f(z) \qquad , \qquad (1D.7)$$

where $f(z)$ contains all the longitudinal dependence and $g(\rho)\cos(n\phi+\psi)$ contains the transverse dependence. Then the coupling coefficient for a large number of small distortions can be simply expressed,

$$d(p) = |D(p)|^2 < |F(\Delta\beta(p))|^2 > \qquad . \qquad (1D.8)$$

In doing this we must insist that the departure of the waveguide from the ideal symmetry is either small or slow compared with wavelength of the propagating light. The quantity $D(p)$ depends only on the transverse index distortion and has the form,

$$D(p) \sim D_{pp'} \sim \int (\vec{E}_p)^* \cdot (\vec{E}_{p'})g(\rho)\cos(n\phi+\psi)\rho d\rho d\phi \quad , \quad (1D.9)$$

where $\vec{E}$ is the transverse electric field vector for the pth mode group of the particular waveguide type being considered. It is seen that $D_{pp'}$ is a measure of the degree to which the distortion causes an overlap of the fields of the p and p' modes groups. It is this quantity which also determines the selection rules for mode coupling. This quantity may be exactly calculated using the $\cos(n\phi+\psi)$ distortion for the step ($\alpha = \infty$) and the parabolic ($\alpha = 2$) profiles. Using the WKB method for the index profile in Eq. (1A.74) an expression for $D(p)$ may be developed which linearly interpolates between these two results for an arbitrary α,

$$D(\alpha,p) = \frac{nka}{2\sqrt{2}} \left[\frac{p}{P}\right]^{2/\alpha+2} (\Delta\beta)^2 \qquad . \qquad (1D.10)$$

The quantity $F(\Delta\beta(p))$ is the fourier transform of the longitudinal distortion function $f(z)$ over the length of the guide,

$$F(\Delta\beta(p)) = \frac{1}{\sqrt{L}} \int_0^L f(z)e^{-i\Delta\beta z} \, dz \qquad . \qquad (1D.11)$$

If the function $f(z)$ contains no spatial frequencies corresponding to the separation, $\Delta\beta(p) \equiv \beta_p - \beta_p'$ between the p and p' mode groups, then no mode coupling will occur. Obviously a sinusoidal longitudinal dependence will couple only a single mode group spacing. As seen in Eq. (1A.80) the mode group spacing depends on the type of waveguide.

All mode groups of the parabolic guide ($\alpha = 2$) can be coupled with a single frequency, whereas a spectrum of frequencies is required for the step guide ($\alpha = \infty$). If Eq. (1D.10) is integrated twice by parts, it is mathematically equivalent to,

$$F(\Delta\beta) = \frac{1}{(\Delta\beta)^2} \frac{1}{\sqrt{L}} \int_0^L \frac{d^2f}{dz^2} e^{-i\Delta\beta z} dz \qquad . \qquad (1D.12)$$

The quantity d^2f/dz^2 is the curvature of the waveguide axis whose distortion is given by $f(z)$. If $C(\Delta\beta)$ is the fourier transform of this curvature, then,

$$F(\Delta\beta) = \frac{1}{(\Delta\beta)^2} C(\Delta\beta) \qquad . \qquad (1D.13)$$

The quantity $<|F(\Delta\beta)|^2>$ then is seen to be the average power spectrum of either the distortion of the waveguide axis or its curvature. Thus by combining Eq. (1D.8), (1D.9) and (1D.11) we have the coupling coefficient for the case of random curvatures of the waveguide axis. The curvature function is specified by $C(\Delta\beta)$.

Even though the function $d(p)$ can be specified, Eq. (1D.3) can only be solved in terms of standard function only for certain types of distortions. One such form which is fairly general and has physical significance is,

$$d(p) = d_0 \left[\frac{p}{P_c}\right]^{-2q} \qquad . \qquad (1D.14)$$

For $q = 0$ the coupling has the constant value d_0.

<u>Steady state solution.</u> Before considering pulse be-
havior of mode coupling, insight can be obtained by anal-
yzing the case where $\partial P/\partial t = 0$. With the preceding assumptions on $\gamma(p)$ and $d(p)$, Eq. (1D.3) can be written,

$$\frac{\partial P}{\partial z} = - \gamma_0 P + \frac{d_0}{P_c^2} \frac{1}{\chi} \frac{\partial}{\partial \chi} \chi^{1-2q} \frac{\partial P}{\partial \chi} \qquad , \qquad (1D.15)$$

where the normalized group number $\chi = p/P_c$ has been introduced. The solution to this equation is (26),

$$P_j(\chi,z) = P_j(\chi) e^{-(\gamma_0 + \gamma_j)z} \qquad , \qquad (1D.16)$$

where,

$$P_j(\chi) = N_j \chi^q J_\ell(\lambda_j \chi^{1+q}) \qquad , \qquad (1D.17)$$

and

$$\gamma_j = \frac{d_0}{P_c^2}(1+q)^2 \lambda_j^2 \qquad . \qquad (1D.18)$$

The order of the Bessel function is $\ell = \pm q/1+q$. The
boundary condition (1D.5) that no power can exist at $\chi > 1$
gives,

$$J_\ell(\lambda_j) = 0 \qquad . \qquad (1D.19)$$

This condition defines the jth value of λ, which is simply
the jth zero of J_ℓ, $Z_j(\ell)$. The normalization factor N_j is
obtained from the other boundary condition Eq. (1D.6),

$$N_j = (1+q)^{\frac{1}{2}} |J_{\ell+1}(Z_j(\ell))|^{-1} \qquad . \qquad (1D.20)$$

The most general solution to Eq. (1D.3) is a superposition
of solutions in Eq. (1D.16),

$$P(\chi,z) = e^{-\gamma_0 z} \sum_{j=1}^{\infty} I_j P_j(\chi) e^{-\gamma_j z} \qquad (1D.21)$$

where the coefficients I_j are determined from the initial power distribution $I(\chi)$, at $z = 0$,

$$I_j = \int_0^1 2\chi I(\chi) P_j(\chi) d\chi \quad .$$ (1D.22)

Equation (1D.21) shows that the power $P(\chi,z)$ is attenuated by the normal loss coefficient, γ_0. Additionally it is made up of a linear combination of modes whose power distributions $P_j(\chi)$ are each attenuated by an excess loss coefficient, γ_j, which depends on the magnitude of the coupling coefficient, d_0. Since γ_j is proportional to the square of the jth zero of the ℓth Bessel function and since $Z_j(\ell)$ increases with j, it is seen that for large z the term $\exp(-\gamma_j z)$ rapidly attenuates all terms above $j = 1$. Thus the steady state mode coupled power distribution will quickly approach,

$$P(\chi,z) = I_1 P_1(\chi) e^{-(\gamma_0 + \gamma_1)z} \quad .$$ (1D.23)

Thus for a step fiber with constant coupling between mode groups, i.e., $q = 0$, the steady state distribution will be,

$$P(\chi,z) \sim J_0(2.405\chi) \exp\{-d_0 \left[\frac{2.405}{P_c}\right]^2\} \quad .$$ (1D.24)

For a step guide for which $\chi = p/P_c = \theta/\theta_c$, the far field angular distribution will exhibit this distribution.

<u>Time dependent mode coupling.</u> When the proper dependence of $v(p)$ is inserted into Eq. (1D.3) no analytic solution is possible. Even a computer computation of the power distribution for a specific case is a complex problem. However, a great deal of information as well as a considerable simplification can be obtained by considering the moments of the impulse response as defined in Eq. (1C.5) and (1C.6).

By introducing the Laplace transform $R(\chi,z,s)$,

$$R(\chi,z,s) = \int_0^\infty P(\chi,z,t) e^{-st} dt \quad ,$$ (1D.25)

Eq. (1D.32) may be transformed to,

$$\frac{\partial R}{\partial z} + \frac{s}{v(\chi)} R = -\gamma_0 R + \frac{d}{P_c^2} \frac{1}{\chi} \frac{\partial}{\partial \chi} \chi^{1-2q} \frac{\partial R}{\partial \chi} \quad .$$ (1D.26)

The moments, σ_n of the distribution R, may be obtained from Eq. (1C.6) by replacing $\omega \to s$ and $P \to R$.

This equation is very reminiscent of Eq. (1D.15) and suggests that the solution for R is a linear combination of the steady state solutions already found in Eq. (1D.16),

$$R(\chi,z,s) = \sum_{j=1}^{\infty} a_j(z,s)P_j(\chi)e^{-(\gamma_0+\gamma_j+sn/c)z} \quad . \qquad (1D.27)$$

The exponential factor sn/c is arbitrarily inserted to null out the delay common to all modes. When Eq. (1D.27) is inserted into Eq. (1D.26) the differential equation for the coefficients a_j is obtained,

$$\frac{\partial a_j}{\partial z} + s \sum_{j=1}^{\infty} M_{jk}a_j e^{-(\gamma_j-\gamma_k)z} = 0 \quad , \qquad (1D.28)$$

where,

$$M_{jk} = \int_0^1 2\chi \left(\frac{1}{v(\chi)} - \frac{n}{c}\right) P_j(\chi)P_k(\chi)d\chi \quad . \qquad (1D.29)$$

From the value of $v(\chi)$ (Eq. (1C.22)) to first order in Δ the matrix M_{jk} is simply,

$$M_{jk} = \frac{n\Delta}{c} \left(\frac{\alpha-2-\epsilon}{\alpha+2}\right) \int_0^1 2\chi^{\frac{3\alpha+2}{\alpha+2}} P_j(\chi)P_k(\chi)d\chi \quad . \qquad (1D.30)$$

By using an iterative perturbation technique to solve Eq. (1D.28) the pulse moments are found in terms of the quantity M_{jk}. The first three have been evaluated (26) and we examine the results.

The zeroth moment, σ_0 is the total power passing the point z. In the limit that $z > 1/\gamma_1$, it simply becomes the steady state distribution already given in Eq. (1D.23).

The first moment or mean delay time in this limit is,

$$T(z) \simeq \left(\frac{n}{c} + M_{11}\right) z + C_T \quad , \qquad (1D.31)$$

where C_T is a constant depending on differences in the initial power distribution. As expected the mean arrival increases linearly with z. The quantity M_{11} determines the effective velocity of the mode coupled distribution.

49

Finally the asymptotic $(z > 1/\gamma_1)$ form for the rms pulse width is,

$$\sigma(z) = \sqrt{2z} \ \{\sum_{j=2}^{\infty} \frac{\left|M_{j1}\right|^2}{\gamma_j - \gamma_1} \}^{\frac{1}{2}} \ . \tag{1D.32}$$

This equation demonstrates the beneficial effect of mode coupling on the information carrying capacity of a waveguide. Without coupling, the rms width increases directly with guide length at a rate given by Eqs. (1C.24) and (1C.25). However, with coupling the rms width increases proportional to $\sqrt{\text{length}}$ and hence the rate of decrease of information capacity with length is considerably less.

From the form of M_{jk} in Eq. (1D.30) it is seen that to first order in Δ, $\sigma(z)$ in the presence of mode coupling exhibits the same minimum at $\alpha = 2 + \epsilon$ as obtained for the optimal gradient in Eq. (1C.24) but now with a considerably different length dependence.

Attenuation with mode coupling. We next consider the loss associated with this mode coupling. To do this, as shown by Eqs. (1D.8) and (1D.13) the curvature spectrum must be known. In general this is not possible. However, Olshansky (27) has found that for the case of a fiber passing over a large number of bumps as in Fig. (D.1) and subject to an elastic restoring force, the curvature power spectrum can be generalized to,

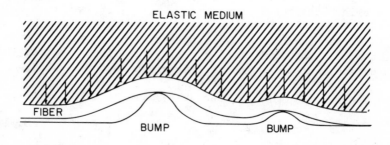

Fig. D.1. Schematic diagram of fiber configuration for which the distortion power spectrum can be calculated.

$$<|c(\Delta\beta)|^2> = C_0(\Delta\beta D)^{-2\eta} \quad , \qquad (1D.33)$$

where C_0 characterizes the strength of the coupling, D measures the correlation length of the curvature and η is an arbitrary parameter characterizing the power spectrum. The case $\eta = 0$ corresponds to a constant curvature power spectrum. Combining Eqs. (1D.8), (1D.10), (1D.13), (1D.14) and (1D.33) and substituting them into Eq. (1D.18) gives the excess loss due to mode coupling produced by this power spectrum,

$$\gamma_1 = \frac{c(\alpha,\eta)}{\Delta} \left(\frac{a^2}{\Delta}\right)^\eta \quad , \qquad (1D.34)$$

where

$$c(\alpha,\eta) = \frac{C_0}{2} \frac{1}{D^{2\eta}} \left(\frac{\alpha+2}{4\alpha}\right)^{1+\eta} (1+q)^2 Z_1^2(\ell) \left(\frac{P}{P_c}\right)^{2(1+q)} \quad . \quad (1D.35)$$

The dependence on the index gradient is contained in Eq. (1D.35) and is plotted in Fig. D.2) for three values of η. For $\eta = 2$ there is a two-fold increase in C in going from a step ($\alpha = \infty$) to a parabolic ($\alpha = 2$) guide, while for $\eta = 0$ little difference in C between the two guide types exists.

The behavior of the loss with respect to the core radius and the relative index difference bears examination. It is noted that the loss due to random curvatures decrease as Δ increases and as the core radius decreases. This would suggest easy remedies for these type losses. However, tradeoffs obviously exist in that as Δ increases the bandwidth decreases (Eq. (1C.24)) and as the core radius decreases the coupling efficiency is reduced. A careful balance for a particular situation is therefore required.

There is a single parameter which can be used to characterize mode coupling quite well. From Eq. (1D.32) the rms mode coupled pulse width can be put in the form,

$$\sigma(z) = \sigma_c(\alpha,\eta) \sqrt{\frac{z}{\gamma_1}} \quad , \qquad (1D.36)$$

where ,

51

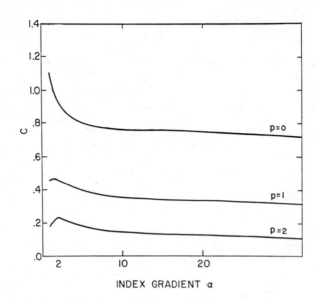

Fig D.2. Plot of the steady state mode coupling attenuation coefficient as a function of index gradient α for three different types of distortion. (From R. Olshansky, Appl. Opt.,14, 941 (1975),reproduced with permission.)

$$\sigma_c = \{2Z_1{}^2 \sum_{\ell=2}^{\infty} \frac{|M_{1\ell}|^2}{Z_\ell{}^2 - z_1{}^2}\}^{\frac{1}{2}} \quad .$$ (1D.37)

A plot of $\sigma(z)$ is shown in Fig. D.3 for a guide with 4 dB/km of normal loss, $\Delta = 0.01$ and a distortion spectrum $\eta = 1$. It is seen that, depending on the level of excess loss, it makes a smooth transition between two asymptotic forms,

$$\sigma(z) = \sigma_u z \qquad\qquad z \ll 1/\gamma_1 \quad ,$$

and   (1D.38)

$$\sigma(z) = \sigma_c \sqrt{\frac{z}{\gamma_1}} \qquad\qquad z \gg 1/\gamma_1 \quad ,$$

where σ_u is obtained from Eqs. (1C.24) and (1C.25). The intersection between these two points is called the coupling length L_c,

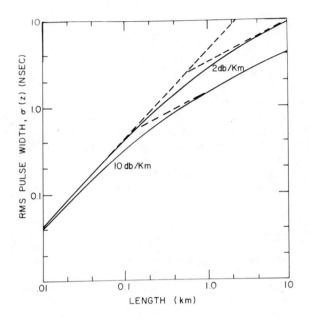

Fig. D.3.　RMS pulse width as a function of length for two different levels of mode coupling attenuation. (From R. Olshansky, Appl. Opt., 14, 943, (1975), reproduced with permission.)

$$L_c = \frac{1}{\gamma_1} \left(\frac{\sigma_c}{\sigma_u}\right)^2 \qquad . \qquad (1D.39)$$

The total loss in the case of mode coupling is,

$$\beta(dB) = 4.3 \, \gamma_1 z \qquad . \qquad (1D.40)$$

If a quantity R is constructed, which is the ratio of the mode coupled to the uncoupled rms pulse width, then it is seen that,

$$R^2\beta = 4.3 \left(\frac{\sigma_c}{\sigma_u}\right)^2 = 4.3 \, \gamma_1 L_c = \text{constant.} \qquad (1D.41)$$

Thus $R^2\beta$ is recognized as the mode coupling loss per unit coupling length. It depends only on α, η, Δ and on the initial modal power distribution but is independent of the fiber length. Because of this it gives a good measure of the tradeoff between the increased bandwidth and the decreased transmission due to mode coupling. This quantity is plotted in Fig. D.4 as a function of α for a series of η values. It is seen to exhibit a value of approximately 0.5 for the step guide, $\alpha = \infty$. The discontinuity near $\alpha = 2$ results because the sharp dip in $\sigma(z)$ occurs at slightly different α values with and without mode coupling. Therefore R experiences a rather large change. Since practical waveguides may not have exact α-profiles, the near parabolic guide is expected to have $R^2\beta \sim 2$. Thus to achieve the same relative bandwidth increase for the near parabolic guide, a loss penalty $\sim 4X$ greater than for the step guide will be incurred.

As the final consideration in the area of mode coupling, fiber buffering is considered briefly. In the process of packaging waveguides for practical applications, great care must be used to avoid incurring losses due to random bending as just discussed. It has been shown (27,28) that by

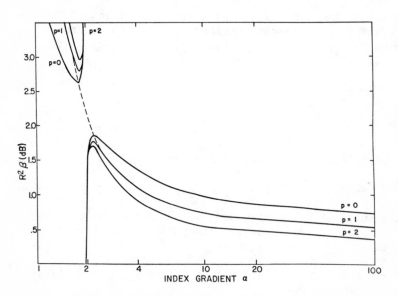

Fig. D.4. Mode coupling parameter, $R^2\beta$ as a function index gradient for three types of distortion. (From R. Olshansky, Appl. Opt., 14, 944 (1975), reproduced with permission.)

sheathing the waveguide in a low modulus material, this loss can be reduced. Considering the model shown in Fig. D.1 and using the theory for random bending, the loss has been evaluated to be (27),

$$\gamma = N<h^2> \frac{a^4}{b^6 \Delta^3} \left(\frac{E}{E_f}\right)^{3/2} \quad , \quad (1D.42)$$

where N is the number of bumps of average height h per unit length, b is the total fiber diameter, a is the core radius and E_f and E are the elastic moduli of the fiber and the fiber surrounding respectively. Thus it is seen that since E_f can be 3 orders of magnitude larger than E for many plastic materials, this type of shielding provides a means of reducing losses due to mode coupling. Additionally this loss is seen to decrease strongly with increasing fiber size which furnishes another design parameter.

The propagation characteristics of fiber waveguides which have been considered in this work are nicely summarized in Fig. D.5 in a plot of the anticipated fiber bit rate as a function of fiber length. This is plotted assuming a GaAs laser with 1 mwatt coupled into the fiber, a $\Delta = 0.01$, a Si-APD detector operating at optimal gain assuming an error rate of 10^{-9}, an $R^2\beta = 0.5$, and a 5% deviation from the optimal index gradient. Material dispersion bounds the plot on the upper right. At some length, loss in the waveguide will produce a signal limited condition as shown by the near vertical lines. Within the remaining region, systems may be considered. As shown, mode coupling can enhance the transmission bandwidth, being traded off, however, against having to decrease the overall fiber loss or the system length.

An attempt has been made to cover in some detail the operation of fiber optical waveguides. The reader is referred to the original works for more complete information. At the present time the future looks very bright for a vast variety of systems to use the unique capabilities of optical waveguides. General concepts of the type presented here will provide the base upon which future progress will be built.

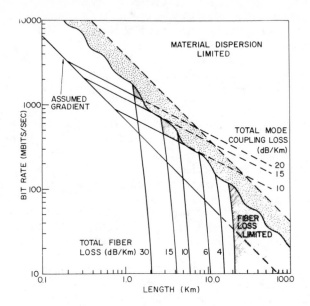

Fig. D.5.   Transmission bit rate as a function of fiber
length.

References

1.   E. G. Rawson, D. R. Herriott and J. McKenna, "Analysis
     of Refractive Index Distributions in Cylindrical,
     Graded-Index Glass Rods (GRIN Rods) Used as Image
     Relays," Appl. Opt., 9, 753 (1970).

2.   D. Marcuse, "Light Transmission Optics," p. 10, Van
     Nostrand Reinhold Company, New York, 1972.

3.   A. W. Snyder, "Asymptotic Expressions for Eigenfunctions
     and Eigenvalues of a Dielectric Optical Waveguide,"
     IEEE Trans. Microwave Theory Tech., MTT-17, 1310 (1969).

4.   D. Gloge, "Weakly Guiding Fibers," Appl. Opt., Vol. 10,
     2252 (1971).

5. D. Marcuse, "Theory of Dielectric Waveguides," p. 72, Academic Press, New York, 1974.

6. E. Merzbacker, "Quantum Mechanics," Chap. 7, John Wiley and Sons, New York, 1961.

7. C. N. Kurtz and W. Streifer, "Guided Waves in Inhomogeneous Focussing Media," IEEE Trans. Microwave Theory Tech., MTT-17, 250 (1969).

8. D. Gloge and E. A. J. Marcatili, "Multimode Theory of Graded Core Fibers," Bell Syst. Tech. J., 52, 1563, (1973).

9. D. B. Keck, "Spatial and Temporal Power Transfer Measurements on a Low Loss Optical Waveguide," Appl. Opt., 13, 1882 (1974).

10. D. B. Keck, R. D. Maurer, and P. C. Schultz, "On the Ultimate Lower Limit of Attenuation in Glass Optical Waveguides," Appl. Phys. Let., 22, 307 (1973).

11. D. S. Carson and R. D. Maurer, "Optical Attenuation in Titania Silica Glasses," J. Non-Cryst. Solids, 11, 368, (1973).

12. R. D. Maurer, E. J. Schiel, S. Kronenberg and R. A. Lux, "Effect of Neutron and Gamma-Radiation on Glass Optical Waveguides," Appl. Opt.,12, 2024 (1973).

13. I. D. Aggerwal, P. B. Macedo, and C. J. Montrose, "Light Scattering in Lithium Aluminosilicate Glass System," American Ceramic Society Meeting, Bedford Springs, Pa., 1974.

14. R. H. Stolen, E. P. Ippen, and A. R. Tynes, "Raman Oscillation in Glass Optical Waveguides," Appl. Phys. Let., 20, 62 (1972).

15. R. G. Smith, "Optical Power Handling Capacity of Low Loss Optical Fibers as Determined by Stimulated Raman and Brillouin Scattering," Appl. Opt., 11, 2489 (1972).

16. J. D. Crow, "Power Handling Capability of Glass Fiber Lightguides," Appl. Opt., 13, 467 (1974).

17. D. Gloge, "Propagation Effects in Optical Fibers," IEEE Trans. Microwave Theory Tech., MTT-23, 106 (1975).

18. C. Pask, A. W. Snyder and D. J. Mitchell, "Number of Modes on Optical Waveguides," J. Opt. Soc. Am., 65, 356 (1975).

19. E. A. J. Marcatili and S. E. Miller, "Improved Relations Describing Directional Control in Electromagnetic Wave Guidance," Bell Syst. Tech. J., 48, 2161 (1969).

20. F. P. Kapron, D. B. Keck and R. D. Maurer, "Radiation Losses in Glass Optical Waveguides," Appl. Phys. Let., 17, 423 (1970).

21. D. B. Keck and R. D. Maurer, "Research on Glass Wave-guides for Optical Communications," 1st European Electro-optics Markets and Technology Conference and Exhibition, Geneva, Switzerland, 1972.

22. S. D. Personick, "Receiver Design for Digital Fiber Optic Communication Systems," Bell Syst. Tech. J., 52, 843 (1973).

23. R. Olshansky and D. B. Keck, "Material Effects on Minimizing Pulse Broadening," Topical Meeting on Optical Fiber Transmission, Williamsburg, Va., 1975.

24. D. Gloge, "Optical Power Flow in Multimode Fibers," Bell Syst. Tech. J., 51, 1767 (1972).

25. D. Marcuse, "Theory of Dielectric Waveguides," Chap. 3, Academic Press, New York, 1974.

26. R. Olshansky, "Mode Coupling Effects in Graded-Index Optical Fibers," Appl. Opt., 14, 935 (1975).

27. R. Olshansky, "Model of Distortion Losses in Cabled Optical Fibers," Appl. Opt., 14, 20 (1975).

28. W. B. Gardner and D. Gloge, "Microbending Loss in Coated and Uncoated Optical Fibers," Topical Meeting on Optical Fiber Transmission, Williamsburg, Va. 1975.

Chapter 2 - OPTICAL FIBER CABLE

James E. Goell

ITT Electro-Optical Products Division

Roanoke, Virginia

2.1 INTRODUCTION

Optical fibers offer numerous advantages such as small size, wide bandwidth, and low attenuation to communication systems of the future. However, before the inherent advantages of optical fibers can be realized, suitable means must be found to package them for practical use.

As drawn, optical fibers are extremely strong; however, abrasion and chemical attack can seriously reduce their strength. Furthermore, the cabling process can introduce bends which can cause optical radiation and thus increase optical attenuation. Plastic coating techniques have been developed to preserve fiber strength and to inhibit excess cable loss. By using these techniques, special cable designs and special cabling techniques, it has been shown that strong, lightweight, small size, handleable, low attenuation cables can be produced. This chapter describes the principles governing mechanical and optical performance of optical fiber cables, gives some early cable designs, and describes some of the results that have been obtained to date.

Optical fiber cable development is now in progress at ITT Electro-Optical Products Division, ITT Standard Telecommunications Laboratories, Corning Glass Works, Bell Telephone Laboratories, several laboratories in Japan, and at many other organizations. Much of the work is proprietary, and so is not generally accessible. The principal works on low loss cables for which information is available were done at ITT and Corning under contracts with the Naval Electronics Laboratory Center and the Army Electronics Command, respectively, and it is these efforts from which most of the results reported on here are taken. It is to be expected that the performance of optical cables will improve rapidly as a result of the effort now in progress, and so the work presented here should be viewed as preliminary in nature.

2.2 MECHANICAL CONSIDERATIONS

The key considerations which must be addressed in the development of optical fiber cables are fiber strength, cabling induced stresses, and use of load carrying members.

2.2.1 Fiber Strength

The tensile strength of pristine glass fiber is comparable to that of any material including steel. However, mechanical damage by drawing equipment, exposure to the atmosphere, and handling leads to a severe deterioration of fiber strength. In particular, the formation of microcracks results in a substantial reduction of tensile strength. Optical properties may also be adversely affected. In order to produce a strong fiber for cabling, the fiber must be either coated in line to protect it, after drawing to strengthen it, or both. The coating may also serve to inhibit the increase in optical attenuation (excess loss) introduced in the cabling process.

A variety of plastic materials and coating techniques exist which are applicable to the protection of optical fibers. Prime factors to be considered are:

o Protection against mechanical damage

o Protection against moisture

o Processability and compatibility with fiber manufacturing

o Compatibility with the jacketing process

A number of materials have physical properties which suggest their use as fiber coatings. Fluoropolymers exhibit low water vapor permeability. Some, such as Kynar[R] can be dissolved and applied from solution while others, such as Teflon[R] FEP and PFA must be applied by extrusion. Polyurethane can also be applied by extrusion. Many other potential candidates also exist.

Alternate methods to dipcoating and extrusion are spray coating and electrostatic coating. With suitable effort it is anticipated that all of the above processes can be employed satisfactorily for some applications.

In the Corning-Army cable program an initial thin coat of Kynar was dipcoated on the fiber. Following this, a thicker layer of plastic was extruded to inhibit excess cable loss.

For the ITT-Navy program in-line extrusion was employed without a primary dipcoat.

2.2.2 Cabling Induced Stress

Conductors are often incorporated into a cable by wrapping them around a central strength member or around each other, which introduces stresses. Three stresses—tension, torsion, and flexure—must be considered when designing an optical cable.

If the fibers are tightly wrapped around a core of strength members or other fibers, the fibers will be in tension. Such tension must be kept to a safe level. The ultimate allowable level is yet to be determined and will depend on progress in producing strong fiber, allowable yield, and design life.

Wrapping the fibers around a core can also induce torsion. Figure 2.1 shows a typical machine used to cable conventional conductors with low

Fig. 2.1   Planetary Stranding Machine
(Krupp Industrie-Und Stahlbau)

torsion. The large wheel of the machine rotates as the cable is drawn through it. For low lay angles (small pitch) this would introduce one rotation of the fiber for each wrap around the core. To compensate for twist to reduce conductor torsion, a series of gears and chains are incorporated behind the wheel so that as it rotates the wire spool axes maintain a fixed orientation.

Since the fiber is bent around the strength member, flexural stress is induced. The flexural stress is given by

$$\sigma_f = \frac{ED}{2R}$$

$$R = R_m \left[1 + \left(\frac{P}{2\pi R_m} \right)^2 \right]$$

where R = bending radius of the fiber

E = modulus of elasticity

d = diameter of fiber

$$R_m = \frac{D+d}{2}$$

D = diameter of center core plus wall thickness of coating

P = pitch or lay

Figure 2.2 gives curves showing the fiber bend radius and flexural stress as a function of helix pitch for a cable with the following characteristics:

CENTRAL STRENGTH MEMBERED FIBER OPTIC CABLE

E = 9×10^6 psi

d = .005"

R_m = .125"

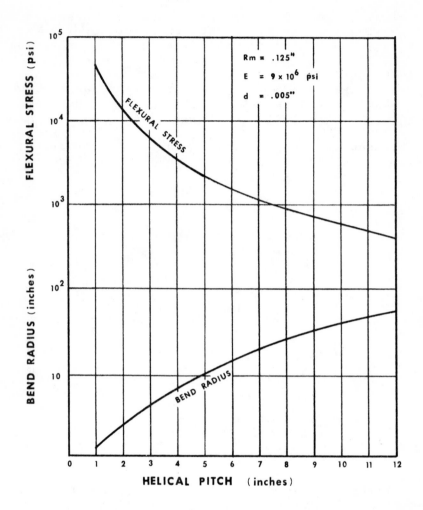

Fig. 2.2   Bend Radius and Flexural Stress vs
Helical Pitch

These curves show that when the pitch increases the flexural stress decreases and the bend radius increases. At a 5" lay, the flexural stress is about 2200 psi. This value is very low compared to the anticipated minimum tensile strength of the fiber.

2.2.3 Strengthening

For applications requiring high tensile strengths - above about 25 lbs. - strength members will have to be incorporated in the cable.

To achieve high tensile strength, materials for members must have both a high tensile strength and a high Young's modulus. In addition, the choice of materials for strength members also depends on the required bending radius of the cable, permissible stress levels to which the core may be subjected during cable manufacture and allowable fiber elongation. For example, if strength members and fibers are layed longitudinally, the strain of the strength member under load must be less than the ultimate strain of the fiber. Furthermore, to achieve high flexibility multi-filament yarn strength members must be used in contrast to strength members made from solid materials. Among materials which promise to meet these requirements are Kevlar Yarn, and steel. The stress-strain curves of these materials are shown in Figure 2.3, along with that of a typical glass fiber for comparison.

Kevlar is a high modulus material which can be obtained as a multi-filament yarn coated with polyurethane. The coating increases the strength and life of the Kevlar bundle by reducing friction between filaments and holding broken filaments in place. Kevlar is available in two grades, a high modulus grade marketed under the trade name Kevlar 49 and a high tensile strength grade marketed under the trade name Kevlar 29. As shown in Figure 2.3, Kevlar 49 exhibits a lower percentage elongation than Kevlar 29 under identical load conditions, and it is, therefore, better suited for application

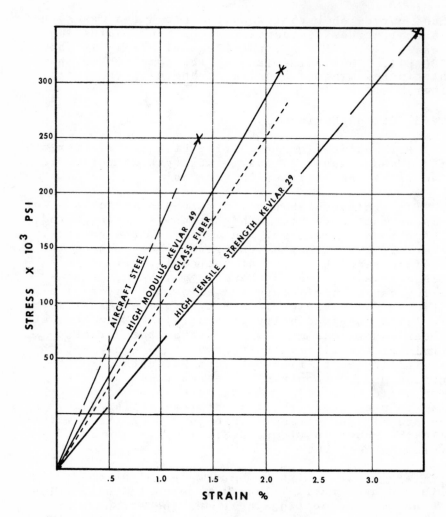

Fig. 2.3 Stress vs Strain for Steel and
Kevlar Strength Members

as a strength member material in fiber optic cables.
Steel has a higher Young's modulus than Kevlar, and
it has superior tensile properties to Kevlar unless
glass fibers with ultimate elongations above about
1.8 per over long lengths can be developed. However,
for applications requiring a non-conducting strength
member steel cannot be used.

2.3 OPTICAL CONSIDERATIONS

2.3.1 Excess Cable Attenuation

Within a cable, stress applied perpendicular
to a fiber's axis can induce small radius bends if
structural irregularities are present. Axial stress
can also introduce bends when irregularities are
present and can elongate the fiber and open microcracks. The effect of these phenomena on fiber
attenuation must be taken into account in the design of cables and in the selection of fibers for
a particular application.

Bend induced radiation can significantly increase the loss of cabled fibers. Generally, a
radius of curvature of $R_{min} = r_c/(NA)^2$ will result
in a very high transmission loss, where r_c is the
fiber core radius and NA the fiber numerical aperture. The fiber should have a curvature significantly greater than r_c. For large radii of
curvature, R, the effect will decrease exponentially
according to $\exp(-R/R_{min})$.

Since the fiber core is already under tension
due to differential cooling and dissimilarity in
coefficients of expansion and fictive temperatures
of the glasses employed, direct tension is not
expected to affect the loss. Tension over a bump
can cause a bend and induce radiation. Tension can
also open microcracks and cause stress induced refractive index changes. Whether this last effect
can become significant before the fiber fails is in
question. In any case, it is expected that work
aimed at improving fiber strength will reduce the
microcrack density and make this question academic.

Figure 2.4 shows the case of fiber subjected to transverse stress over a bump in a cable. According to an early theory*, the dependence of the distortion loss on the core radius r_c and the relative index difference Δ is given by

$$\alpha_b = \frac{cr_c^4}{\Delta^3} \qquad (2\text{-}1)$$

where $c = 0.9 \, p \frac{h^2}{b^6} \left(\frac{E_m}{E_s}\right)^{3/2}$

α_b = distortion loss

r_c = core radius

b = fiber diameter

Δ = relative index difference $1 - \dfrac{n_{core}}{n_{clad}}$

h = effective rms bump height

p = number of bumps per unit length

E_m = modulus of encapsulating material

E_s = modulus of fiber core

With the assumed values

r_c = 2.0 mils
b = 5.0 mils
Δ = 0.0053   (NA = 0.15)
E_m = 10^5 psi
E_s = 9×10^6 psi

*R. Olshansky, "Distortion Losses in Cabled Optical Fibers," Applied Optics, Vol. 14, No. 1, Jan 75, pp 20-21.

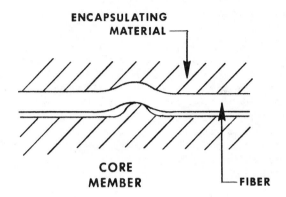

Fig. 2.4   Transverse Stress Induced Bend

the loss would be 0.0045 dB for each .1 mil bump.

An examination of Eq. 2-1 shows that excess cable losses can be reduced considerably by increasing the diameter and numerical aperture of fibers and by decreasing the core to cladding ratio.

For the case of a graded index fiber, it is anticipated that the effective numerical aperture for bend loss considerations will be about half the peak numerical aperture, which may necessitate the use of fibers with small cores. The excess loss of single mode fibers has not been fully analyzed. Assuming that Eq. 2-1 can be applied to single mode operation, the application of single mode fibers would be extremely attractive due to their small core, high numerical aperture, and low dispersion.

A plastic coating can serve to stiffen a fiber, and thus to reduce microbend loss. It has been shown that thick plastic coatings can be applied by melt extrusion without seriously increasing attenuation. In the Corning-Army cable program the loss

69

before and after extrusion coating was compared.
Table I gives some of the comparisons. In the ITT-
Navy program losses as low as 4 dB/km at .85u were
achieved. Before and after results cannot be given
for the ITT work because the plastic was applied on-
line.

C. Kao and G. Bickel of ITT have performed
experiments to evaluate the effect of stress on the
attenuation of fibers. For all of their tests .23
numerical aperture, 50 um core diameters, 125 μm
cladding diameter Kynar coated CVD fibers were
employed. The purpose of the tests was to simulate
the effects of elongation, tension over bumps,
transverse stress induced bending, and bends induced
by residual stresses in the fiber coating. The
tests and their results were as follows:

1. The fiber was wound on a drum and the drum heat-
    ed producing a 0.2 percent increase in drum cir-
    cumference. The effect of elongation, and
    tension induced bends caused by fiber coating
    and drum irregularities were evaluated. No
    change in attenuation was observed over the full
    range of the test.

2. The fiber was multiply wound between two drums
    and the fibers strained to 0.2 percent by separ-
    ating the drums to observe the same phenomenon
    as in 1. Again no attenuation change was
    observed.

3. A 0.01 inch diameter rod was inserted under the
    fibers wound and strained as described in 2 to
    observe tension induced bend loss. Again no
    attenuation change was observed.

4. The fibers wound as in 2 were interwoven with
    three 0.01 inch diameter rods to observe the
    effect of transverse stress induced bending on
    fiber loss. For this test a significant incre-
    ase was measured as shown in Figure 2.5.

The results in tests 1, 2 and 3 are in marked
contrast of some of the results that have been performed by other workers. However, in those tests,
fibers with larger cores and lower numerical

TABLE I

EFFECT OF EXTRUSION COATINGS
ON ATTENUATION

LUBRICANT + BASE COATING	COATING	ATTENU- ATION @ 820 nm in dB/km	NUMER- ICAL APER- TURE
Silane (Z6079) +Kynar^R	NONE	6.2	.134
Silane (Z6079) + Kynar^R	POLYURETHANE (Estane^R 58300)	16.4	.134
Silane (Z6079) + Kynar^R	NONE	5.4	.205
Silane (Z6079) + Kynar^R	POLYURETHANE (Estane^R 58300)	5.2	.205
Silane (Z6079) + Kynar^R	NONE	6.8	.196
Silane (Z6079) + Kynar^R	POLYURETHANE (Roylar^R E-9) Loose	6.1	.196
Silane (Z6079 + Kynar^R	NONE	6.5	.154
Silane (Z6079) Kynar^R	POLYURETHANE (Roylar^R E-9) Tight	8.0	.154

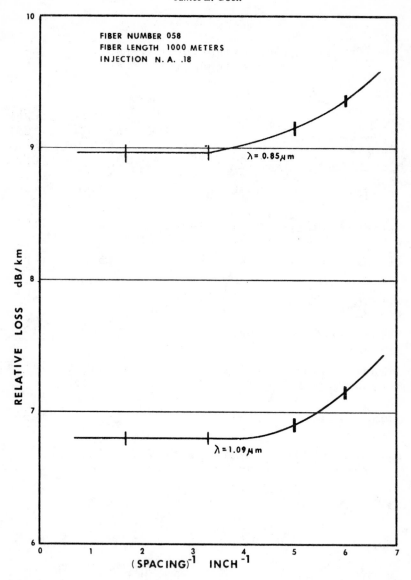

Fig. 2.5 Loss Increase Caused by Induced Bend

apertures were used. Thus, from the standpoint of
excess loss, the use of small core, high numerical
aperture fibers is advantageous.

2.3.2 Transmission Characteristics

Both attenuation--length dependent loss-- and
end dependent losses contribute to the insertion
loss of an optical cable. Attenuation is comprised
of both the intrinsic fiber attenuation α_i and the
excess cable attenuation α_e. The one time losses
consist of fresnel reflections at both ends of the
cable, I_f; and the light gathering efficiency I_s.
The total insertion loss is given by $10 \log I = (\alpha_i + \alpha_e)L + 20 \log I_f + 10 \log I_s$ dB. The fresnel
loss is given by

$$I_f = \frac{(n_c - n_e)}{(n_c + n_e)}$$

The gathering efficiency which increases with
numerical aperture and source-fiber common area,
will not be dealt with in detail here.

For very short cables, the end losses make the
dominant contribution to the total insertion loss
while for very long cables the attenuation will
dominate. Thus, for some short distance applications with large area sources, it is advantageous
to use a number of fibers in parallel while for long
cables it is doubtful that the advantage obtained
will justify the added expenses.

2.4 CABLE DESIGNS

The cable design should be chosen with full
consideration of the end use. For example, the
operating conditions for end loaded cables for tow
or sonabuoy applications are radically different to
those for a pole mounted cable. For a ducted cable
the operating conditions are usually mild, but the
installation conditions may be severe. For com-
puter interconnects the main problem may be instal-
ler care while for aircraft applications the

mechanical stress may be small, but the temperature
constraints severe.

A variety of cable designs have been proposed
and others will emerge with time. At present, those
of most interest include simple bundles of high loss
high numerical aperture fibers, smaller bundles of
low loss high strength fibers, ribbon cable, and
strengthened cables.

Figures 2.6a and 2.6b show central and peripheral strength membered cable designs, which are the
two basic cylindrical configurations. The flat
cable shown in Figure 2.7 is also of interest. Such
cables tend to be strong, and have high flexure resistance on one plane. The "feel" of the cable in
the other plane suggests care in bending, but use by
untrained personnel accustomed to making folds with
copper conductored cables may be a problem.

2.4.1 Bundles

The simplest fiber optic cables structure consists of a group of fibers collected in a bundle and
surrounded with a plastic jacket. Generally, the
term bundle has implied that the fibers will be used
in parallel.

Bundle cables can be broken into two types.
One with a large number of high loss fibers (typically 200-400 with high numerical aperture), and
the other with a smaller number (7-37) of low loss
fibers.

High loss fibers are presently relatively in-
expensive and thus generally are the only ones which
can be used in large numbers. Large bundles are
presently used with large area sources so the large
number of fibers is advantageous from the standpoint
of collection efficiency. Since breakage can be
large, it also improves reliability. Typical bundles of this type have attenuations between 400
dB/km and 2000 dB/km and cables with significantly
lower losses are possible. The numerical aperture
of the fibers is generally between .5 and .6.

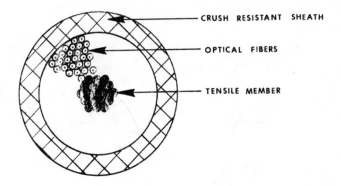

Fig. 2.6a   Central Strength Member Configuration

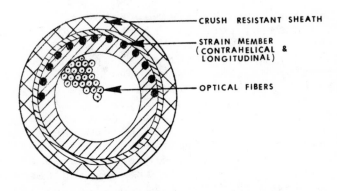

Fig. 2.6b   Peripheral Strength Member
            Configuration

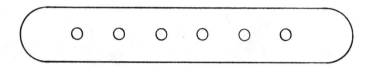

Fig. 2.7 Ribbon Cable

These cables are best suited for short runs where
their collection efficiency makes them attractive.

Low loss bundles are now reaching the market.
One such bundle, produced by the Corning Glass Works,
contains 19 fibers with a minimum attenuation of
about 20 dB/km, and a numerical aperture of .15.
The diameter of the fibers cores is 85 μ and the
overall fiber diameter is 125 μ. The outside diameter of the bundle jacket is 3.3mm. A high performance version of the cable has also been
described by Corning. A 500 meter length of the
cable had 90% of the fibers intact. The typical
attenuation of the cable at .82 μ was 15 dB/km. At
any wavelength, the bundle attenuation was at least
3 dB/km larger than the original attenuation of any
fiber in the cable.

2.4.2 Strengthened Cable Designs

The Army and Navy sponsored cable programs have
yielded cables for which experimental results are
now available. The Army-Corning program resulted in

76

the external strength membered design shown in
Figure 2.8 and the Navy program the internal strength membered design shown in Figure 2.9. The Corning
cable consists of 7 Kynar coated fibers with sili-
cone oil lubricated polyurethane jackets as pre-
viously described. The center fiber is provided as
a space filler to allow helical fiber laying. The
six fibers for the ITT cable were extrusion coated
with Teflon^R FEP directly on the draw tower.

Both cables have been tested for attenuation
and tensile strength. The Corning cable was also
subjected to MIL-C-13777F impact, twist and bend
tests, while MIL-C-13777F tests on the ITT cable
have just begun.

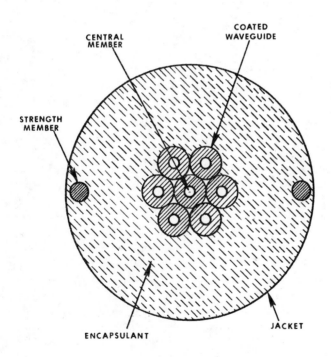

Fig. 2.8 Corning Cable Design

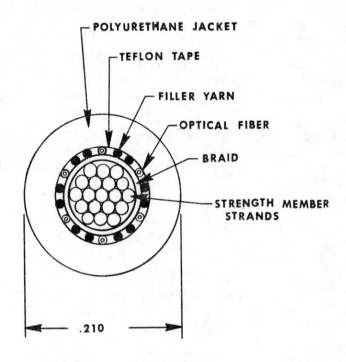

Fig. 2.9 ITT Cable Design

Figure 2.10 shows the attenuations of the best and worst fibers in the Corning cable as a function of wavelength. Similar data for the best ITT cable is shown in Figure 2.11. The differences in performances are due to differences in the characteristics of the fibers chosen for the cables.

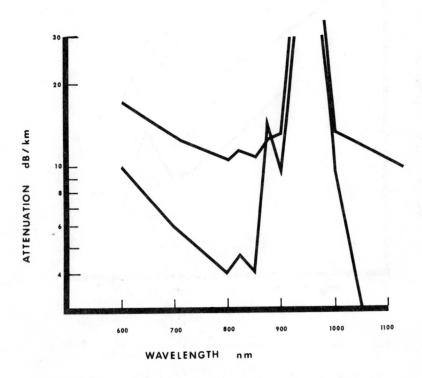

Fig. 2.10   Cabled Fiber Attenuation vs. Wavelength for
the Corning Optical Fiber Cable (Lowest
and Highest Attenuation Fibers)

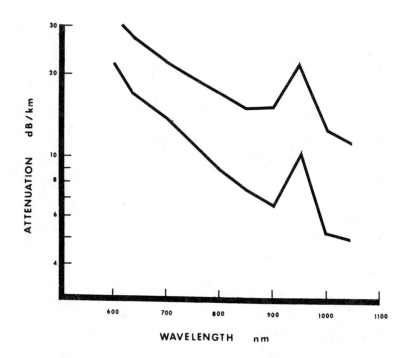

Fig. 2.11 Cabled Fiber Attenuation vs. Wavelength for
the ITT Optical Fiber Cable (Lowest and
Highest Attenuation Fibers

The Corning cable was tested for tensile strength using a 2 ft. gage length. In half of their tests, initial fiber failure occurred at a 175 lb. load. A 5 meter gage length of the ITT cable was subjected to a 500 lb. load without breakage. For a 20 meter gage length 5 of 6 fibers survived a 500 lb. load. The difference in strength is accounted for by the cross sectional area of Kevlar employed.

The Corning cable withstood over 200 impacts at 1 ft-lb and 2 out of 4 cables withstood 200 impacts at 1.5 ft-lb before the first fiber broke. An average of 56 impacts were required to break the first fiber at 1.75 ft-lb. In a single test, 57 2.5 ft-lb impacts were required to break the first fiber with the ITT cable and 209 the second.

The differences in the results obtained in these initial military optical cable development programs resulted in part from differences in objective and with additional effort it is anticipated that both approaches can be improved. Thus, it is not possible to draw general conclusions about the relative merits of the two approaches. The above data is significant because it shows that optical cables can be produced which are suitable for practical field use.

ACKNOWLEDGMENT

Much of the work described here is freely quoted from previous publications and direct communications. The author wishes to acknowledge R. A. Miller of Corning, M. Pomerantz of ECOM, R. L. Eastley of NELC, A. Asam, M. Maklad, J. C. Smith, G. Bickel, and C. Kao of ITT whose contributions to cable technology served as a basis for this chapter. He also acknowledges the assistance of R. W. Dillon in preparing the manuscript.

CHAPTER 3 - COUPLING COMPONENTS FOR OPTICAL
FIBER WAVEGUIDES

M. K. Barnoski

Hughes Research Laboratories
Malibu, California 90265

I. INTRODUCTION

Recent development of low-loss glass fiber waveguides
has substantially increased the interest in optical communications. As a result, the variety of optical fiber
waveguides which are becoming available is also increasing.
These fibers have varying attenuations, numerical apertures,
and diameters, and are packaged in various ways ranging
from simple bundles containing as many as 5,000 fibers to
actual cables containing only six or seven fibers. This
large array of available fibers and the large assortment
of light-emitting and laser diodes now commercially available pose the question of what source to select, given a
fiber or fiber bundle. It is the intent of this chapter to
present an elementary review of coupling the radiation
emitted by the source into the fiber waveguide.

2. OPTICAL COUPLING INTO GLASS FIBER
MULTIMODE WAVEGUIDES

2.1 Power Transfer Between Emitting and
Receiving Surfaces

An estimate of the input coupling efficiency of the
source to the fiber can be obtained by considering the
power transfer efficiency between a radiating surface of
area A_S and a receiving surface of area A_R. The power

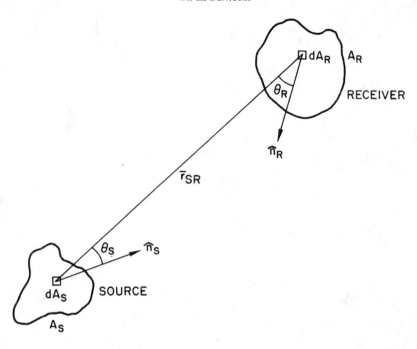

Fig. 3.1. Arbitrary orientation of emitting and receiving
surfaces.

transfer can be computed if the goniometric characteristics
of the two surfaces are known. Consider the two arbitrarily
oriented surfaces separated by distance vector $\bar{r}_{SR}$, as
shown in Fig. 3.1. The power transfer from an element of
area dA_S to an element of area $\cos\theta_R\ dA_R$ normal to r_{SR} is

$$dP_{SR} = B_\Omega(\bar{X}_S,\theta_S)\ dA_S\ d\Omega_R, \tag{3.1}$$

where $B_\Omega(\bar{X}_S,\theta_S)$ is the brightness of the source in units of
watts per square centimeter steradian. For sources with
nonuniform radiation distributions, B_Ω is a function of $\bar{X}_S$
which is the position vector on the surface A_S. The
quantity $d\Omega_R$ is the solid angle subtended by $(\cos\theta_R\ dA_R)$ at
the source point dA_S, that is,

$$d\Omega_R = \frac{\cos\theta_R\ dA_R}{r_{SR}^2}. \tag{3.2}$$

The total power transfer between A_S and A_R is, therefore,

$$P_{SR} = \int_{A_S} \int_{A_R} B_\Omega(\overline{X}_S, \theta_S) \cos\theta_R \, r_{SR}^{-2} \, dA_R \, dA_S. \qquad (3.3)$$

In general, this integral is complicated with the angles θ_S and θ_R and the separation vector r_{SR} being dependent on the position coordinates on the surfaces A_R and A_S. In the interest of obtaining an estimate of the input coupling efficiency, approximations can be made which considerably simplify the integration.

Reference to the two arbitrarily oriented differential surface areas in Fig. 3.2 reveals that the area $dA_R \cos\theta_R$, which is normal to the vector $\overline{r}_{SR}$ is approximately equal to the differential area of a sphere centered at dA_S, i.e., $dA_R \cos\theta_R \simeq r_{SR}^2 \sin\theta_S \, d\theta_S \, d\phi$. The angle ϕ is an angle varying from 0 to 2π in the source plane. The total power transferred between A_S and A_R is, therefore,

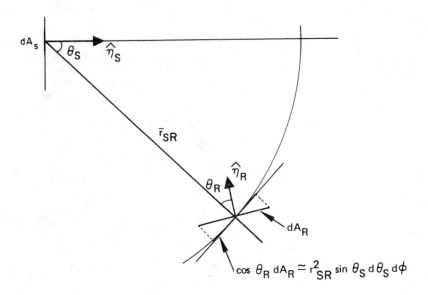

Fig. 3.2 Arbitrarily oriented differential surface areas, dA_S and dA_R.

$$P_{SR} = \int_{A_S} \int_{\theta_S} \int_{\phi} B_{\Omega}(\overline{X}_s, \theta_s) \sin\theta_s \, d\theta_s \, d\phi \, dA_S. \quad (3.4)$$

The integration is considerably simplified if the source has a uniform radiation distribution across its area and if the extent of the source area is small compared with the separation vector r_{SR}. When these conditions are satisfied, the total power transfer becomes

$$P_{SR} = 2\pi A_S \int B(\theta) \sin\theta \, d\theta, \quad (3.5)$$

where cylindrical symmetry has also been assumed.

2.2 Goniometric Characteristics of Opto-Electronic Sources

There are two sources of interest for use in fiber optic communications systems — the GaAs light-emitting diode and the GaAs injection laser diode. Although light-emitting diodes (LED's) are currently available in a variety of geometrical configurations for the applications considered here, they can be catagorized into two basic types. These are the small area, high brightness emitters currently under development and the more readily available, large area, low brightness devices. The optical emission areas of com-mercially available, low brightness LED's typically range from 0.25 mm^2 to 5.1 mm^2. These source areas are considera-bly larger than the core area of low-loss, multimode fiber. A typical core diameter for a low-loss fiber is 85 μm. As a result of the large area mismatch, a large coupling loss is incurred which, when added to the losses resulting from numerical aperture effects, makes the utilization of these large area emitters impractical for a fiber link using a single or only a few strands as a transmission channel. They do, however, have applications in systems employing fiber bundles whose diameter is equal to or greater than that of the source.

In addition to the low brightness LED's, there are state-of-the-art, high brightness LED's and cw, room temperature, injection lasers now becoming available on a limited basis. The emission areas of these devices are well matched to the core area of a single fiber. High brightness, light-emitting diodes can be flat geometry surface emitters such as those developed by Burrus[1] or strip geometry edge emitters under development at RCA laboratories.[2] A cross-sectional drawing of a double hetero-junction electroluminescent diode coupled to a single optical fiber is shown in Fig. 4.16 of Chapter 4. The 50 µm diameter emission area of this device is well matched to the core area of low-loss fiber.

The angular distribution of radiation emitted from the surface of such an LED can be determined with the aid of the schematic diagram shown in Fig. 3.3.

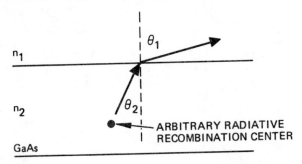

Fig. 3.3 Planar-type GaAs LED schematic.

The radiation from any arbitrary point interior to the LED is isotropic into the full 4π steradians; however, due to refraction at the interface of the LED whose refractive index is n_2 and the surrounding medium of index n_1, the radiation emerging from a planar geometry LED has an angular dependence.

Let the photometric intensity be I(watts/steradian). The conservation of energy requires that

$$I_2 \, d\Omega_2 = I_1(\theta_1) \, d\Omega_1 \tag{3.6}$$

or

$$I_2 \sin\theta_2 \, d\theta_2 \, d\phi = I_1(\theta_1) \sin\theta_1 \, d\theta_1 \, d\phi, \tag{3.7}$$

which, using Snell's law of refraction

$$n_2 \sin\theta_2 = n_1 \sin\theta_1 \tag{3.8}$$

and its differential

87

$$n_2 \cos\theta_2 \, d\theta_2 = n_1 \cos\theta_1 \, d\theta_1, \tag{3.9}$$

yields

$$I_1(\theta_1) = \left(\frac{n_1}{n_2}\right)^2 I_2 \frac{\cos\theta_1}{\cos\theta_2}. \tag{3.10}$$

Expressing θ_2 in terms of θ_1 and substituting into equation (3.10) results in

$$I_1(\theta_1) = \left(\frac{n_1}{n_2}\right)^2 I_2 \frac{\cos\theta_1}{\left[\left(1 - \left(\frac{n_1}{n_2}\right)^2 \sin\theta_1^2\right)\right]^{1/2}}, \tag{3.11}$$

which since

$$\frac{n_1}{n_2} < 1, \tag{3.12}$$

becomes approximately

$$I(\theta) \simeq \left[\left(\frac{n_1}{n_2}\right)^2 I_2\right] \cos\theta. \tag{3.13}$$

The external radiant intensity from a planar geometry surface-emitting LED is the same as a planar radiating surface with brightness of value

$$B(\theta) = B \cos\theta \tag{3.14}$$

with

$$B = \left[\left(\frac{n_1}{n_2}\right)^2 \frac{I_2}{A_{source}}\right] \tag{3.15}$$

where A_{source} is the emitting area of the LED and I_2 depends on drive current, quantum efficiency, and reflectivity. For ranges of θ of interest, the internal reflectivity is approximately constant; hence I_2 can be considered constant. The radiation pattern obtained from a 50 μm active diameter, surface-emitting LED manufactured by Plessey Opto

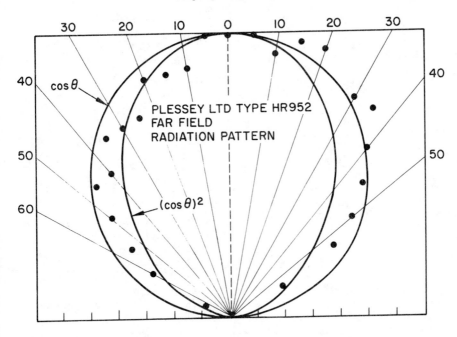

Fig. 3.4. Angular radiation pattern of Plessey HR952 LED.

Electronics is shown in Fig. 3.4. As can be seen, the
distribution approximates a $\cos\theta$ distribution as was
predicted for a planar radiating surface. A surface which
radiates with a $\cos\theta$ distribution is termed a Lambertian
radiator.

A more directional beam can be obtained from an edge-
emitting, double heterojunction, light-emitting diode.
This results from the fact that the active p-n junction
region of the device is sandwiched between two layers of
semiconductor material with refractive index less than that
of the active region. The optical waveguiding which results
yields a more directional external radiation pattern.
Typical radiation patterns obtained from an edge-emitting,
electroluminescent diode are shown in Fig. 4.14 (Chapter
4). Also shown in the figure are the radiation patterns
obtained from a double heterojunction laser diode. As can
be seen, the emission pattern obtained from the laser diode
is much more well directed than that of the LED.

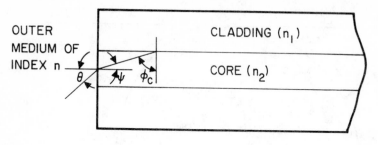

$$+n \sin \theta = n_2 \sin \psi = n_2 \sin(90 - \phi_c) = (n_2^2 - n_1^2)^{1/2}$$

Fig. 3.5. Entrance angle of cladded fiber for total
internal reflection at core-cladding interface.

2.3. Acceptance Angle of the Fiber

The amount of energy coupled into a fiber is greatly
dependent on its numerical aperture (NA) which, for a step
index fiber, can be defined with the aid of Fig. 3.5. For
total internal reflection to occur, the internal reflection
angle must be greater than the critical angle $\sin\phi_c = n_1/n_2$.
This can be related to the incident angle θ via Snell's law

$$+ n \sin\theta = n_2 \sin\psi = n_2 \sin(90 - \phi_c) = (n_2^2 - n_1^2)^{1/2}.$$

The quantity $n \sin\theta$ is termed the numerical aperture
(NA) of the fiber. For the case when the outer medium is
air, NA $= \sin\theta \simeq \theta$ for small NA. Since the fiber accepts
only those rays contained within a cone whose maximum angle
is determined by total internal reflection at the core-clad
interface, an input coupling loss will result if the angular
emission cone of the source exceeds that defined by the
numerical aperture of the fiber. This is illustrated
schematically in Fig. 3.6 for a source coupled to a single
strand. Only those rays contained within the shaded cone
are trapped in the core.

The optical ray illustrated in Fig. 3.5 is a meridional
ray, that is, a ray which passes through the axis of the
waveguide at some part in its path. In addition to meridional rays there are also skew rays which do not intersect
the fiber axis. A plane wave incident on the core of the
fiber excites both meridional and skew rays. Meridional

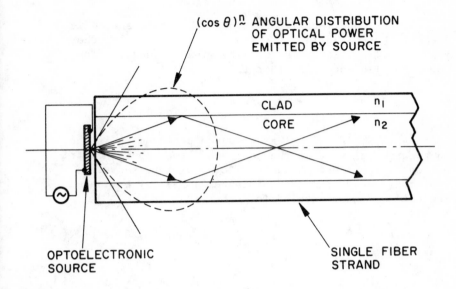

$$P_{FIBER} \cong P_{SOURCE} \left(\frac{n+1}{2}\right) N.A.^2$$

$$N.A. = (n_2^2 - n_1^2)^{1/2}$$

Fig. 3.6. Schematic diagram of source coupled to single-strand fiber waveguide.

rays are not excited if the irradiance angle θ exceeds the numerical aperture of the fiber. Skew rays, however, can be excited beyond the numerical aperture of the step index fiber.[3] These high angle rays are often neglected because they are usually highly attenuated. The power carried by skew rays will be neglected in the following analysis of coupling efficiency.

2.4. Coupling Efficiency

Direct Coupling. If the opto-electronic source is directly butted against the fiber or fiber bundle, the input coupling factor can be determined directly. The power coupled into the fiber is

$$P_{fiber} = 2\pi\, A_{source}\, f_P \int_o^{\theta NA} B(\theta)\, \sin\theta\, d\theta, \qquad (3.16)$$

where A_{source} is the emitting source area and f_P is the packing fraction. For the case where the angular distribution of the source brightness can be expressed as $B(\theta) = B(\cos\theta)^n$, the above integral is easily evaluated. The optical power coupled into the fiber is

$$P_{fiber} = 2\pi\, A_{source}\, f_P\, B \int_o^{\theta NA} (\cos\theta)^n\, \sin\theta\, d\theta \qquad (3.17)$$

or

$$= 2\pi\, A_{source}\, B\, f_P \left[\frac{1-(\cos\theta_{NA})^{n+1}}{n+1} \right]. \qquad (3.18)$$

The total optical output power emitted from the source into the full hemisphere is given by

$$P_{source} = 2\pi\, A_{source}\, B \int_o^{\pi/2} (\cos\theta)^n\, \sin\theta\, d\theta \qquad (3.19)$$

or

$$= 2\pi\, A_{source} \frac{B}{n+1}. \qquad (3.20)$$

The power coupled into the fiber can, therefore, be expressed in terms of the total optical power emitted from the source, that is,

$$P_{fiber} = P_{source}\, f_P \left[1-(\cos\theta_{NA})^{n+1} \right] \qquad (3.21)$$

which for small NA reduces to

$$P_{fiber} = P_{source}\, f_P \left(\frac{n+1}{2} \right) NA^2 \quad \text{(for small NA)}. \qquad (3.22)$$

It should be noted that the assumption has been made that

the cross-sectional area of the radiation pattern at the
entrance plane of the fiber is less than or equal to the
fiber bundle area (i.e., no area mismatch loss). For
coupling into a single fiber strand the packing fraction
is, of course, unity.

Once again this analysis includes only the bound or
trapped rays which travel in the direction of the waveguide axis and which undergo total internal reflection at
the core-clad interface. It neglects the leaky modes
which travel obliquely (skew) to the waveguide axis and
undergo only partial reflection at the core boundary. The
leaky modes are suitable approximations for the radiation
field within the fiber. In some cases a significant
portion of the radiation field can persist for long distances
in the fiber waveguide. A complete analysis must, therefore, include the leaky rays. C. Pask and A. W. Snyder[4]
treat this problem in detail using a modified form of
geometric optics. The interested reader is referred to
Ref. 4.

Coupling Using a Lens. The above results obtained
with the source placed directly against the fiber or fiber
bundle revealed that the input coupling coefficient was
proportional to the square of the numerical aperture of
the fiber. Since the numerical aperture is less than unity
(typical values range from 0.14 to 0.5), a considerable
amount of power emitted by the source can be lost in input
coupling. It is of interest to investigate the effect on
the coupling loss of introducing an intervening optical
element between the radiating source and fiber bundle, as
illustrated in Fig. 3.7. For simplicity, it is assumed
that the radiation emanating from the source is a constant
within the solid angle Ω_s, that is, the angular distribution
of the source brightness is assumed to be of the form

$$B(\theta) = \begin{cases} B & \theta \leq \theta_s \\ 0 & \theta > \theta_s \end{cases} \qquad (3.22)$$

Again for simplicity, cylindrical symmetry is assumed. A
detailed treatment of coupling both disc and strip geometry
devices with and without a lens where these simplifying
approximations are not made can be found in Ref. 5.

93

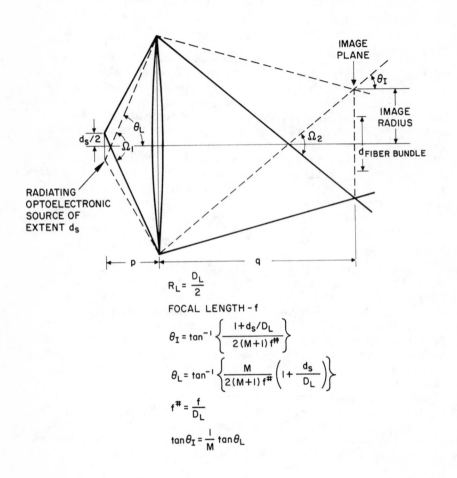

$$R_L = \frac{D_L}{2}$$

FOCAL LENGTH – f

$$\theta_I = \tan^{-1}\left\{\frac{1+d_s/D_L}{2(M+1)f^{\#}}\right\}$$

$$\theta_L = \tan^{-1}\left\{\frac{M}{2(M+1)f^{\#}}\left(1+\frac{d_s}{D_L}\right)\right\}$$

$$f^{\#} = \frac{f}{D_L}$$

$$\tan\theta_I = \frac{1}{M}\tan\theta_L$$

Fig. 3.7. Geometry used for lens input coupler calculations.

The total power collected by the lens, that is, the total power in the input cone Ω_1, is

$$P_{lens} = 2\pi \ B \ A_{source} \int_o^{\theta_L} \sin\theta \ d\theta \qquad (3.23)$$

$$P_{lens} = 2\pi \ B \ A_{source}(1 - \cos\theta_L) \qquad (3.24)$$

where

$$\theta_L = \tan^{-1}\left\{ \frac{M}{2(M + 1) \ f^{\#}} \left(1 + \frac{d_s}{D_L}\right)\right\} \qquad (3.25)$$

and again for simplicity, it has been assumed that the thin lens formula applies (i.e., $1/p + 1/q = 1/f$). The definitions for the quantities, $M, f^{\#}, D_L$ are

M = magnification = q/p

D_L = lens diameter

f = focal length of lens

$f^{\#}$ = f-number of lens = f/D_L.

The formulas developed herein involving the trigonometric expressions, including those shown in Figure 3.7, are exact to the extent that the thin lens formula applies.

The angular distribution in the input cone Ω_1 is uniform, provided that the lens collection angle θ_L is less than or equal to the maximum emission angle θ_s from the source, that is, $\theta_L \leq \theta_s$. For this case, the power collected by the lens must also be uniformly distributed in the output cone Ω_2 where

$$\Omega_2 = 2\pi \int_o^{\theta_I} \sin\theta \ d\theta = 2\pi(1 - \cos\theta_I) \qquad (3.26)$$

with

$$\theta_I = \tan^{-1}\left(\frac{1 + \dfrac{d_s}{D_L}}{2(M+1) \ f^{\#}}\right). \qquad (3.27)$$

The watts/steradian contained in the output cone Ω_2 is, therefore,

$$\frac{P_{lens}}{\Omega_2} = B \ A_{source} \frac{(1 - \cos\theta_L)}{(1 - \cos\theta_I)}. \qquad (3.28)$$

The area of the source image (in the image plane of the lens) is $A_I = M^2 A_{source}$. If this image area is less than that of the fiber bundle, that is if

$$M \leq \frac{d_{fiber}}{d_{source}} \equiv M_A, \qquad (3.29)$$

then all the rays in the cone Ω_2 with angles less than the numerical aperture will be collected by the fiber. The total power in the fiber is, therefore,

$$P_{fiber} = 2\pi \frac{P_{lens}}{\Omega_2} \int_0^{NA} \sin\theta \, d\theta$$

$$= 2\pi B f_p A_{source} \frac{1 - \cos\theta_L}{1 - \cos\theta_I} (1 - \cos\theta_{NA}).$$

$$(3.30)$$

If the magnification is such that $M > M_A$, then the above expression must be multiplied by $A_{fiber}/A_{image} = \left(\frac{M_A}{M}\right)^2$.

Now, the total power radiated from the source into Ω_o steradians is given by

$$P_{source} = 2\pi B A_{source} \int_0^{\theta_s} \sin\theta \, d\theta \qquad (3.31)$$

$$P_{source} = 2\pi B A_{source} (1 - \cos\theta_s) \qquad (3.32)$$

where Ω_o is the solid angle which contains the useful external emission being radiated from the emitting diode.

The total optical power coupled into the fiber bundle with an intervening lens is, therefore,

$$P_{fiber} = f_p P_{source} \frac{(1-\cos\theta_L)}{(1-\cos\theta_s)} \frac{(1-\cos\theta_{NA})}{(1-\cos\theta_I)} g_A(M) \quad (3.33)$$

where

$$g_A(M) = \begin{cases} 1 & M \leq M_A \\ (M_A/M)^2 & M > M_A \end{cases} \qquad (3.34)$$

The above calculations clearly reveal that the input coupling loss can be reduced to the packing fraction loss if

the source geometry, bundle configuration, and lens system
are so selected so that

$$\theta_s \leq \theta_L$$

$$\theta_I \leq \theta_{NA}$$

$$M \leq d_{fiber\ bundle}/d_{source}$$

are all satisfied simultaneously. If it is assumed that
the source configuration and lens diameter have been
selected so that $\theta_s = \theta_L$, then

$$P_{fiber} = f_p\ P_{source}\ \frac{(1 - \cos\theta_{NA})}{(1 - \cos\theta_I)}\ \left(\frac{d_{fiber}}{M\ d_{source}}\right)^2 \quad (3.35)$$

which for small angles is approximately given by

$$P_{fiber} \simeq f_p\ P_{source}\ \left(\frac{\theta_{NA}}{1/M\ \theta_s}\right)^2\ \left(\frac{d_{fiber}}{M\ d_{source}}\right)^2 \quad (3.36)$$

since, in the small angle approximation, $\theta_I \simeq 1/M\ \theta_s$.
Clearly, optimized input coupling occurs when $\theta_s = M\theta_{NA}$,
$d_{fiber} = Md_{source}$. The input coupling loss can be reduced
to the packing fraction loss if the angular emission from
the source (θ_s) can be sufficiently collimated so as to have
no angle (θ_I) greater than the numerical aperture of the
fiber, while simultaneously maintaining the cross-sectional
area of the radiation pattern at the input to the bundle
less than the bundle area itself. The results are an
illustration of the law of brightness[6] which states that
the brightness of the image cannot exceed that of the object,
and can only be equal to it if the losses of light within
the optical system are negligible.

Coupling losses equal to the packing fraction losses
have been observed by E. Schiel[7] for a linear, six-strand
bundle of Corning low-loss fiber; an RCA large optical
cavity, whose emission was contained within a ±30° cone;
and a ball lens. The components were arranged as shown in
Fig. 3.8. For these particular source-to-lens and lens-to-
bundle spacings, the magnification of the system is M = 5.
Since the emission area of the source is 2 x 150 $(\mu m)^2$
with a maximum angular output of $\theta_s = 30°$ and since the
linear, six-strand fiber bundle size is 125 x 6 = 750 μm

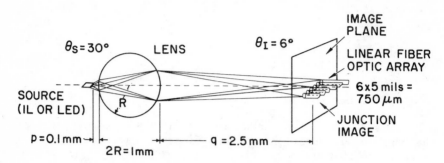

$\theta_S = 30°$ LENS $\theta_I = 6°$

IMAGE PLANE

LINEAR FIBER OPTIC ARRAY

6x5 mils = 750 μm

SOURCE (IL OR LED)

R

JUNCTION IMAGE

P = 0.1 mm

2R = 1mm

q = 2.5mm

MAGNIFICATION: M = 5   SOURCE WIDTH: 150 μm
                        IMAGE WIDTH: 750 μm

Fig. 3.8.   Schematic diagram of image formation by a "ball
lens" source-to-fiber bundle input connector
(courtesy of E. Schiel, U. S. Army Electronics
Command, Ft. Monmouth, N. J.).

with an NA of 6°, an intervening lens with M = 5 reduces
the angular content of the optical radiation to the NA of
the fiber while simultaneously maintaining the size of the
radiation pattern at the input to the bundle equal to that
of the bundle area itself.

2.5.   Sensitivity of Input Coupling to
        Mechanical Alignments

In the design and development of practical input
(source-to-fiber) coupling connectors for single-strand
systems, it is not only of importance to realize the expected
magnitude of the coupling coefficient, but also the effects
of mechanical alignment tolerances on the input coupling
loss. That severe mechanical alignment tolerances are
imposed can be seen from the plot shown in Fig. 3.9 of the
increase input coupling loss as a function of radial
displacement of the center of a Corning low-loss step index
(NA = 0.14) fiber and the center of the 50 μm diameter
surface-emitting LED manufactured by Plessey. As can be
seen from the figure, the input coupling is extremely

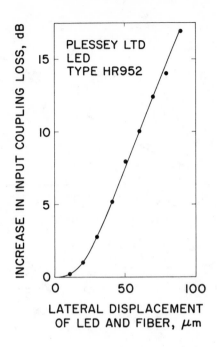

Fig. 3.9.   Increase in input coupling loss as function of
lateral displacement.

sensitive to lateral misorientation.   For example, to
maintain less than a 1 dB increase requires transverse
alignment tolerances of less than ±20 μm.   Note also that
a ±50 μm misalignment of the center of the fiber with
respect to the center of the LED results in a 8 dB increase
in input coupling loss.   The effects on input coupling loss
of longitudinal fiber-to-source separation and angular misalignment of the axis of the source to that of the fiber
are shown in Figs. 3.10 and 3.11.   As can be seen, the
input coupling is relatively insensitive to LED–fiber
separation.   For example, with the LED and fiber separated
by 150 μm, the increase in coupling loss is less than
1.0 dB, while an angular tilt of the fiber axis with respect
to the LED surface of 10° increases the loss by approximately
0.25 dB.   This is consistent with the Lambertian characteristics of the radiation pattern.   The data shown in Figs.
3.9, 3.10 and 3.11 indicate, as expected, that the coupling

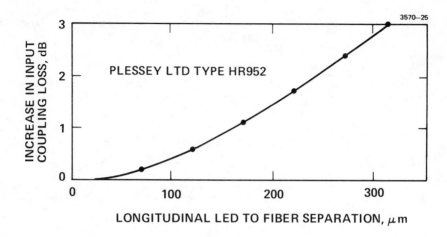

Fig. 3.10.   Increase in input coupling loss as function of
longitudinal LED-to-fiber separation.

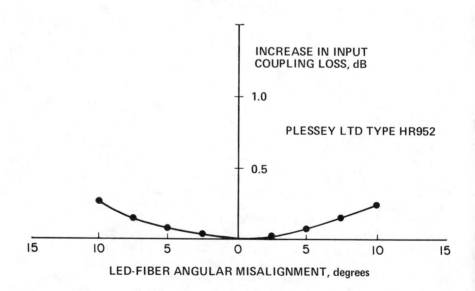

Fig. 3.11   Increase in input coupling loss as a function of
angular misalignment.

loss for the surface-emitting LED is most sensitive to radial misalignment of the LED and fiber core.

Severe mechanical alignment tolerances are not restricted to input couplers using single strands. The mechanical alignment tolerances for the bundle coupler shown schematically in Fig. 3.8 were also found to be very severe.[7]

3. SPLICE COUPLING

In addition to considering the interconnection of the electroluminescent sources and receiving photodiodes with the fiber waveguides, attention must also be given to the problem of splicing the optical fiber cables. Practical system considerations will require that both low insertion loss connect-disconnect splice couplers and permanent fused splice coupling be developed. The quick connect-disconnect type of coupler will find application in the interconnection of fiber cables, connection of cables to repeaters, and the connection of distribution system components such as serial tee and star couplers which will be used in bidirectional, multiterminal data distribution systems.

On the other hand, environmentally sound, permanent cable splices are required for long distance, low-loss, data transmission lines. The engineering design of these splice coupling components will be dependent on the packaging (cable type) and number of fibers to be interconnected at the splice. Systems which attempt to fully utilize the bandwidth and small physical cross-section of the fiber by using individual fibers as isolated channels, each with a transmitter and receiver, have different splice design requirements than a system which employs a bundle of many fibers as the transmission channel. M. I. Schwartz[8] recently summarized the allowable splice characteristics of different types of splices which one can speculate as being required. The table presented by Schwartz is reproduced here as Table 3.1. The table lists allowable characteristics of various types of splice. A unit is a small package of fibers arranged in a geometrical pattern. A cable splice is the splicing of groups of units to groups of units.

TABLE 3.1

Summary of Allowable Splice Characteristics
for Various Types of Splices[8]

Type Splice	Purpose	Mechanical Strength	Speed of Operation	Simplicity of Operation	Loss dB
Single Fiber — Factory Repair		High	Slow	Complex	<0.25
Single Fiber — Field Repair		Low	Slow	Simple	<0.50
Single Fiber — Quick Connect		Low	Fast	Simple	<1.0
Unit Splice — Factory Repair		High	Slow	Complex	<0.25
Unit Splice — Field Repair		Low	Slow	Simple	<0.50
Unit Splice — Quick Connect		Low	Fast	Simple	<1.0
Cable Splice — Factory Repair		High	Slow	Complex	<0.50
Cable Splice — Field Repair		Medium	Slow	Complex	<1.0
Cable Splice — Quick Connect		Low	Fast	Simple	<1.0

From the analysis of the previous sections it should be clear that if the core areas of the fibers to be interconnected are identical and the surface areas perfectly clean and flat, then if the fibers are brought into intimate optical contact, 100% coupling efficiency can be achieved, provided that the matching fibers are perfectly aligned radially and their longitudinal axes are perfectly parallel.

If, however, the optical axes of the two fibers are radially displaced, some of the core area of the fiber will be aligned with the cladding of the receiving fiber. Similarly, if the optical axes of the two fibers are at an angle to one another, some loss of optical energy results. The insertion losses which result from these two types of misalignment are shown in Figs. 3.12 and 3.13 which were extracted from Ref. 9, where some early work done on in-line connectors is described.

The calculated loss as a function of fiber center-to-center displacement shown in Fig. 3.12 illustrates the strongest alignment tolerances for low insertion loss splice connectors. The experimental plots of coupling loss as a function of angular misalignment shown in Fig. 3.13 were obtained using low numerical aperture (NA = 0.14) fibers. The bundle-to-bundle coupling losses were measured on 61-fiber hexagonal bundles.

It can readily be seen from the figures that alignment must be closely maintained to attain good fiber splicing. Such tolerances have not been achieved with conventional connector hardware; therefore, new connector designs need to be developed. In addition to simple alignment problems, the other aspects which must be considered in producing low-loss splices are the Fresnel reflection losses, end finish preparation, and protection of the joint from the ambient environment. The interconnection must also, of course, have sufficient mechanical strength and integrity.

Fiber connectors with low insertion losses are currently under development. For example, K. Miyazaki et al.[10] have reported single fiber connect-disconnect connectors with an average insertion loss of 0.9 dB for 20 connectors evaluated. Low insertion loss, permanent splice connectors have also recently been reported.[11-13] The results presented indicate that permanent splice connection of various

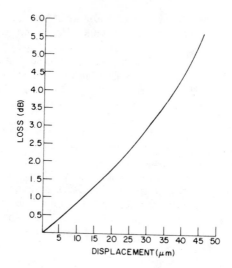

Fig. 3.12. Coupling loss (calculated) as a function of center-to-center displacement for two identical perfectly concentric 75 μm fibers in intimate contact.

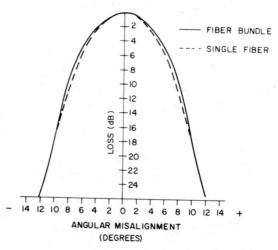

Fig. 3.13. Coupling loss (measured) as a function of angular misalignment for fiber bundles (solid line) and individual fibers (broken line)

104

types of fiber units can be made with average insertion losses in the 0.3 to 0.1 dB range. Although these laboratory results are encouraging, further work needs to be done so that these laboratory techniques can be performed with comparable results in a field environment.

4. SUMMARY

In this chapter an elementary treatment of some of the engineering design considerations that must be considered in the design of input coupling components for fiber optic transmission lines has been presented. A brief discussion splice coupling is also given.

REFERENCES

1. C. A. Burrus and B. I. Miller, Opt. Commun. <u>4</u>, 307, (1971).

2. Refer to Chapter 4.

3. Robert D. Maurer, "Introduction to Optical Waveguide Fibers," in <u>Introduction to Integrated Optics</u>, ed. by M. K. Barnoski (Plenum, New York, 1974), Chapter 8.

4. Colin Pask and Allan W. Snyder, Opto-Electronics <u>6</u>, 297 (1974).

5. K. H. Yang and J. D. Kingsley, Applied Optics <u>14</u>, 268 (1975).

6. M. Born and E. Wolf, <u>Principles of Optics</u> (Pergamon Press, Oxford, 1965), p. 189.

7. E. Schiel, G. Talbot and E. Aras, "Low Loss Coupling of Semiconductor Sources to Multimode Optical Waveguides," presented at Integrated Optics and Fiber Optics Communications Conference, Naval Electronics Laboratory Center, San Diego, Calif., May 15-17, 1974.

8. M. I. Schwartz, "Optical Fiber Cabling and Splicing," in Technical Digest of Topical Meeting on Optical Fiber Transmission, Williamsburg, Va., January 1975.

9. Frank L. Thiel, Roy E. Love and Rex L. Smith, Applied Optics <u>13</u>, 240 (1974).

10. K. Miyazaki, M. Honda, T. Kudo and Y. Kawamura, "Theoretical and Experimental Considerations of Optical Fiber Connectors," in Technical Digest of Topical Meeting on Optical Fiber Transmission, Williamsburg, Va., January 1975.

11. H. Murata, S. Inao and Y. Matsuda, "Connection of Optical Fiber Cable," in Technical Digest of Topical Meeting on Optical Fiber Transmission, Williamsburg, Va., January 1975.

12. E. L. Chennock, D. Gloge, D. L. Bisbee and P. W. Smith, "End Preparation and Splicing of Optical Fiber Ribbons," in Technical Digest of Topical Meeting on Optical Fiber Transmission, Williamsburg, Va., January 1975.

13. A. H. Cherin and P. J. Rich, "A Splice Connector for Joining Linear Arrays of Optical Fibers," in Technical Digest of Topical Meeting on Optical Fiber Transmission, Williamsburg, VA., January 1975.

CHAPTER 4 - ELECTROLUMINESCENT SOURCES FOR FIBER SYSTEMS

H. Kressel

RCA Laboratories

Princeton, N. J. 08540

1. INTRODUCTION

The laser diodes and LED's of primary interest in fiber systems emit in the 0.8-0.9 and 1.0-1.1 μm spectral ranges where the transmission losses are minimal. Source power and modulation requirements depend on the system objective as well as type of fibers used, and the various possible options will not be discussed here. It is evidently desirable, however, to develop a light source with the widest possible system utility, although its full potential would usually not be utilized. For this reason, there has been a long standing interest in a laser diode for optical communications which combines high radiance, ease of direct modulation to GHz rates, small size and potentially low cost. However, it is only since the development of the AlGaAs/GaAs heterojunction laser structures in 1968, [1-3] and important technological progress made after that time, that the potential of laser diodes for fiber communications is being realized.

In parallel with laser diode development, it has been recognized that specially designed LED's could fulfill many system needs in relatively short fiber links with less restrictive operating conditions than those required for lasers. (For example, the temperature dependence of the power output from a laser diode is much greater than from an LED.) The LED requirements for fiber communications have called for the development of special structures capable of

109

reliable high current density operation, a high modulation rate, and a choice of materials to meet the above spectral requirements. Laser diodes and LED's share many aspects of a common technology, and both have benefited from the reliability advances made in recent years.

There is a vast literature concerning all aspects of laser diodes and electroluminescent devices, [4] and the present review is necessarily limited. An introductory treatment concerning spontaneous and stimulated emission in Section 2 is followed in Section 3 by a discussion of the spectral requirements for efficient laser diode operation and the role of heterojunctions. Section 4 is concerned with the materials requirements for the present devices and the role of interfacial defects. State-of-the-art CW laser diode operation is discussed in Section 5, including structural and modal properties. In Section 6, the requirements for the LED's and the major structures used are discussed. Section 7 is concerned with lasers and LED's for 1.0–1.1 μm emission. Finally, major factors affecting diode reliability are reviewed in Section 8.

2. SPONTANEOUS AND STIMULATED EMISSION

2.1 Spontaneous Emission

The electroluminescent diode is basically a device in which electrons and holes are injected into the p- and n-type regions, respectively, by the application of a forward bias V, as shown in Fig. 4.1. The current-voltage characteristics are described by the well known diode equation

$$I = I_s [(\exp qV/a \, kT)-1] \quad , \tag{4.1}$$

where I_s is the saturation current and $1 < a < 2$. The injected minority carriers can recombine either radiatively or nonradiatively. If they recombine radiatively, the emitted photon energy, $h\nu$, is approximately equal to the bandgap energy, E_g, (neglecting recombination via deep centers). For nonradiative recombination, the energy released is dissipated in the form of heat. The two processes are characterized by

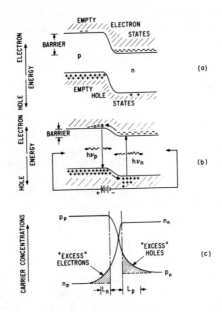

Figure 4.1. Electroluminescent p-n junction operation.
(a) Zero bias. The built-in potential drop across the
p-n junction represents a large barrier for the motion
of teh electrons and holes. (b) Forward Bias. The
potential barrier is significantly reduced by the application of the external voltage. (c) Majority and minority
carrier concentrations on the n- and p-sides of a forward-
biased p-n junction. (Ref. 64)

minority carrier lifetimes, τ_r and τ_{nr}, for radiative and
nonradiative recombination, respectively, which determine
the internal efficiency, η_i, and the lifetime, τ,

$$\eta_i = \frac{1}{1 + \dfrac{\tau_r}{\tau_{nr}}} \, , \tag{4.3}$$

$$\frac{1}{\tau} = \frac{1}{\tau_r} + \frac{1}{\tau_{nr}} \, . \tag{4.4}$$

In the simplest model, the nonradiative lifetime decreases linearly with the density of nonradiative recombination centers, N_t, their capture cross section, σ_t, and the electron thermal velocity, v_{th}

$$\tau_{nr} \approx (N_t \sigma_t v_{th})^{-1} \; . \tag{4.5}$$

The nature of the nonradiative centers is still not well understood in GaAs and related compounds. However, it is generally believed that lattice defects such as vacancies, interstitials, states associated with dislocations, and precipitates of commonly used dopants form nonradiative recombination sites. It is essential, therefore, that the materials used for light-emitting devices be as free from defects as possible.

The reason for the selection of certain semiconductors for LED's and laser diodes is the relationship between the radiative lifetime and the bandstructure. In "direct bandgap" semiconductors electrons and holes across-the-gap from each other have the same value of crystal momentum, making possible direct recombination without a third particle to conserve momentum; this process is very efficient, and the resulting lifetime is very short (10^{-10}-10^{-8} s). The group of direct-bandgap semiconductors includes, among others, GaAs, InAs, InP, $Al_xGa_{1-x}As$ ($0 \leq x \leq 0.37$) and $GaAs_{1-y}P_y$ ($0 \leq y \leq 0.45$). On the other hand, in indirect bandgap semiconductors, such as Si, Ge, GaP, and AlAs, the participation of phonons is required to conserve momentum for across-the-gap electron-hole recombination, and the resultant three-particle recombination process is relatively slow. From the measured absorption coefficient of a semiconductor as a function of photon energy, it is possible to calculate the radiative recombination coefficient, B_r, which determines the radiative minority carrier lifetime for a given majority carrier concentration, N and P, respectively:

$$\tau_r = (B_r N)^{-1} \qquad \text{(n-type material)}$$

$$\tau_r = (B_r P)^{-1} \qquad \text{(p-type material} \tag{4.6}$$

(These above expressions assume that the injected carrier density is substantially below the majority carrier density. If this is not the case, then we enter a regime of bimolecular recombination, and the above expressions overestimate the lifetime.)

Table 4.1 shows the vastly different values of B_r for direct and indirect energy bandgap materials. [5]   In Si, for example, $B_r = 1.79 \times 10^{-15}$ cm^3/s while in GaAs, $B_r = 7.21 \times 10^{-10}$ cm^3/s. For a majority carrier density $P = 10^{17}$ cm^{-3}, the calculated radiative lifetime in Si is $\tau_r = 6 \times 10^{-3}$ s, while in GaAs, $\tau_r = 1.4 \times 10^{-8}$ s. The consequent effect on the internal quantum efficiency becomes obvious when we consider some reasonable values for the nonradiative lifetime. Assuming even a low density of nonradiative centers, say 10^{15} cm^{-3}, with $\sigma_t = 10^{-15}$ cm^2,

TABLE 4.1

CALCULATED RECOMBINATION COEFFICIENT FOR REPRESENTATIVE
DIRECT-AND INDIRECT-BANDGAP SEMICONDUCTORS[a]

Material	Energy Bandgap Type	Recombination Coefficient, B_r (cm^3/sec)
Si	Indirect	1.79×10^{-15}
Ge	Indirect	5.25×10^{-14}
GaP	Indirect	5.37×10^{-14}
GaAs	Direct	7.21×10^{-10}
GaSb	Direct	2.39×10^{-10}
InAs	Direct	8.5×10^{-11}
InSb	Direct	4.58×10^{-11}

(a) Y. P. Varshni, Phys. Stat. Solidi <u>19</u>, 353 (1964).

we obtain an estimate of $\tau_{nr} \approx 10^{-7}$ s. From Eq. 4.3 the calculated internal quantum efficiency would therefore be only 1.7×10^{-5} in Si, but 0.88 in GaAs. A high internal quantum efficiency is clearly much easier to achieve in direct bandgap than in indirect bandgap semiconductors.

2.2   Stimulated Emission

Stimulated emission is achieved by carrier population inversion, a condition where the upper of two electronic levels separated in energy by $E = E_2 - E_1$ has a higher probability of being occupied by an electron than the lower level. The probability of a photon (with energy $h\nu \approx E$) inducing a downward (induced) electron transition will then exceed the probability for an upward transition; i.e., photon absorption. Light amplification becomes possible, therefore, when an incident photon stimulates the emission of a second photon with energy approximately equal to the energy separation of the electronic levels.

The above concepts are general, but semiconductor lasers differ greatly in detail from gas or other types of solid state lasers where the radiative transitions occur between discrete levels of spatially isolated excited atoms. The spontaneous radiation produced by transitions between the isolated atoms extends over a very narrow spectral range while in semiconductors the active atoms are very closely packed causing their energy levels to overlap into bands. The high packing density (about 10^{18} cm^{-3}) of excited atoms in a semiconductor compared to only about 10^{10} cm^{-3} in a gas laser is advantageous because the optical gain coefficient is relatively high, thus allowing for much shorter optical cavities.

Population inversion in a semiconductor is illustrated in Fig. 4.2 which shows the electron energy as a function of the density of states in an undoped semiconductor at a temperature sufficiently low (T = 0 K in the illustration) for the conduction band to be empty of electrons. When electrons are injected, they fill the lower energy states of the conduction band to F_c, the quasi-Fermi level for electrons. An equal density of holes is generated to conserve charge neutrality in the material, and the states in the valence band to F_v are, therefore, empty of electrons.

114

T = O K

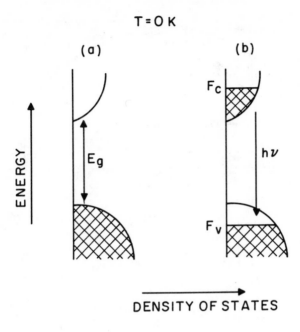

DENSITY OF STATES

Figure 4.2. Electron energy as a function of the density of states in an intrinsic direct bandgap semiconductor at T = 0 K in equilibrium (a) and under high injection (b).

Photons with energy greater than E_g, but less than F_c-F_v cannot be absorbed (since the conduction band states are occupied), but these photons can induce downward electron transitions form the filled conduction band states into the empty valence band states. With increasing temperature, a redistribution of the electrons and holes occurs, which smears out the sharply defined carrier distributions of Fig. 4.2. However, the basic conditions for stimulated emission remain defined as above in terms of the separation of the quasi-Fermi levels, $F_c-F_v > h\nu$.

The requirement for lasing is that the gain matches the optical losses at some photon energy within the spontaneous radiation spectrum. Since the photon density is highest at or near the peak of the spontaneous emission, this is where the gain coefficient is maximum. Figure 4.3a illustrates the position of the lasing peak at threshold

115

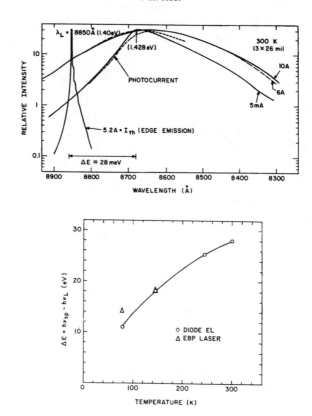

Figure 4.3a. Spontaneous spectra at 300 K below and above lasing threshold as viewed through surface of diode. Lasing spectrum was observed by viewing edge of diode. Photocurrent curve shows dependence of diode photocurrent on incident photon wavelength.

Figure 4.3b. Energy separation between diode spontaneous emission band peak $h\nu_{sp}$ and lasing peak at threshold $h\nu_L$ between 77 and 300 K. Also shown for comparison are two data points for electron-beam-pumped laser (EBP) made from GaAs similar to that in the active region of the heterojunction laser diode. (Ref. 6)

with respect to the spontaneous emission spectrum in a laser diode where both can be observed with no distortion due to internal absorption. [6] The actual peak energy is below the spontaneous emission peak by $\sim$ 0.02-0.03 eV at room temperature for lightly doped material, with the separation decreasing in value with temperature, Fig. 4.3b.

If the semiconductor is highly doped (order 10^{18} cm^{-3} for typical acceptors in GaAs), the radiative transitions involve the impurity states. In the case of p-type dopants (ionization energy $\sim$ 0.03 eV), the emitted photon energy is below the bandgap energy, while for n-type dopants (which have ionization energies < 0.01 eV) the Fermi level shifts upward with doping into the conduction band, and the photon energy exceeds the bandgap energy. If the material is heavily doped and compensated (i.e., contains both donors and acceptors), the lasing photon energy is always below the bandgap energy. The lasing wavelength can be varied from 0.85 to 0.95 μm in GaAs at room temperature with dopant variations, but the device performance is usually best in the 0.88-0.91 μm range.

3. STRUCTURAL REQUIREMENTS FOR EFFICIENT LASER DIODE OPERATION

A unique feature of the laser diode, not present in other laser types, is the ability to obtain stimulated emission by minority carrier injection using a p-n junction or heterojunction. The efficient operation of a laser diode requires effective carrier and radiation confinement to the vicinity of the junction. The average injected electron density from an n-type region into a p-type region (under conditions where the reverse process of hole injection is negligible), is

$$\Delta N \cong \frac{J\tau}{ed} , \tag{4.7}$$

where e is the electron charge, and d is the effective active region width in which the injected average electron density is ΔN.

In a typical GaAs laser diode, the injected carrier
density needed to reach lasing threshold is about 10^{18} cm^{-3}
at room temperature. To minimize the threshold current
density, we restrict the width of the recombination region
d by placing a potential barrier for minority carriers
a distance less than the diffusion length from the p-n
junction. Figure 4.4 shows, for example, how a p-p heterojunction presents a potential barrier for electrons if at
room temperature $\Delta E_g \gtrsim 0.1$ eV. This is the single heterojunction, "close-confinement" laser. [1,2] The use of
a second heterojunction at the injecting p-n junction yields
the double heterojunction [3] diode. It is, however,
essential that the heterojunction interface be relatively
defect-free in order to prevent excessive nonradiative
recombination of the injected carriers (see Section 4).

To ensure wave propagation in the plane of the junction,
means must be provided for stimulated radiation confinement to the region of inverted population (or its close
proximity). An effective way of providing the required
dielectric step(s) for radiation confinement is by the
use of heterojunctions. Two heterojunctions, as indicated
in Fig. 4.5, provide a controlled degree of radiation
confinement due to the higher refractive index in the
vicinity of the recombination region, with the fraction
of the radiation confined dependent on the heterojunction
spacing d and the refractive index steps at the lasing
wavelength.* In general, it is desirable to use a structure
with an equal refractive index step Δn at each heterojunction
to prevent the loss of waveguiding which can occur in
thin asymmetrical waveguides. However, even with the
symmetrical double heterojunction laser, wave confinement
within d is eventually reduced when the heterojunction
spacing becomes too small. [7] This effect is illustrated
in Fig. 4.5, for a relatively wide double heterojunction
structure in which the radiation confinement is nearly
complete (c) and for a very thin heterojunction spacing
in which the optical field intensity spreads symmetrically
on the two sides (e). A controlled degree of radiation
spread is used to obtain devices having a desired far-
field radiation pattern.

---
*In $Al_xGa_{1-x}As$/GaAs structures, $\Delta n \cong 0.62x$ at $\lambda \cong 0.9$ μm.

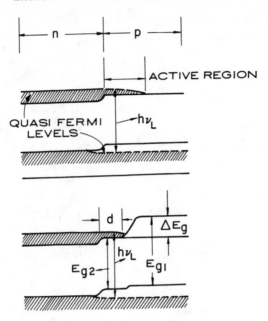

Figure 4.4. Electron distribution in a forward biased homojunction without a potential barrier for carrier confinement (top), and with a p-p heterojunction (bottom) placed distance d less than the diffusion length from the p-n junction (or p-n heterojunction).

The fraction of the radiation confined to the recombination region of the double heterojunction laser is denoted Γ and affects the radiation pattern and threshold current density. The radiation pattern is affected because of the change in effective source size, while the threshold is affected because only the fraction of the optical power within the recombination region is amplified.

The optical feedback is obtained in a laser diode by cleaving two parallel facets to form the mirrors of the Fabry-Perot cavity. The lateral sides of the laser are either formed by roughening the device edges by wire sawing to form a "broad-area" diode, Fig. 4.6a, or by confining the ohmic contact to selected areas to produce

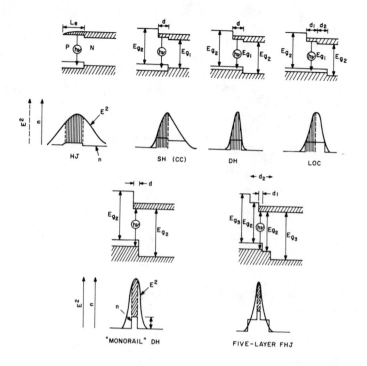

Figure 4.5. Energy diagram, distribution of the refractive index (n) and of the optical energy (E^2), and position of the recombination region in major laser diode structures. (a) Homojunction; (b) single heterojunction (close confinement) diode; (c) double heterojunction with full carrier and radiation confinement; (d) large optical cavity (LOC) diode; (e) double heterojunction with full carrier confinement, but only partial radiation confinement; (f) four-heterojunction diode.

a stripe-contact diode (Fig. 4.6b). This will be discussed further in Section 5.

The conditions for gain in a semiconductor laser cavity, as well as the mathematical expressions for the

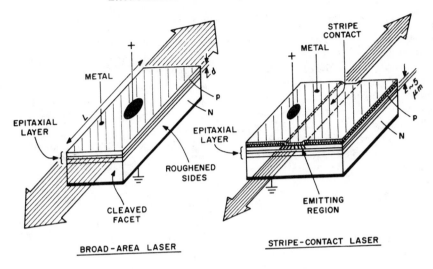

Figure 4.6.   Broad-area and stripe contact laser diodes.

laser threshold current have been treated elsewhere, [4] and will not be considered here. It has been found by many researchers that the threshold current density of typical double heterojunction AlGaAs-GaAs lasers decreases linearly with d, [4f,k]

$$\frac{J_{th}}{d} = (4.0 \pm 0.5) \times 10^3 \, Acm^{-2}/\mu m \, , \qquad (4.10)$$

when $0.3 \leq d \leq 2 \, \mu m$. Hence for $d = 0.3 \, \mu m$, $J_{th} \cong 1200 \, A/cm^2$, a value satisfactory for reliable room temperature CW operation as discussed in Section 5.

In addition to the threshold current density, the differential quantum efficiency η_{ext} above threshold, and the overall power conversion efficiency η_p are important laser diode parameters. In state-of-the-art pulsed laser diodes, η_{ext} = 40-50% at room temperature (emission from both sides). The power conversion efficiency peaks at 2-4 times I_{th} and the best values at room temperature are $\sim$ 22% for two-sided emission, but the power conversion efficiency is only a few percent near threshold.

121

The double heterojunction laser diode is most widely used for CW operation. More complex structures have been extensively studied and these are useful in special applications not requiring CW operation. Figure 4.5d shows the schematic configuration of the LOC [8] (large optical cavity) laser diode in which the p-n junction is bracketed by two heterojunctions, and the four-heterojunction diode [9] where the outer two heterojunctions are added to the two inner heterojunctions to obtain additional control of the optical cavity width. A discussion of these structures is beyond the scope of the present review.

4. MATERIALS CONSIDERATIONS

Table 4.2 shows the III-V compounds with direct bandgap energies of interest in the 0.8-0.9 and 1.0-1.1 μm spectral ranges. Both vapor-phase (VPE) [10] and liquid-phase epitaxy (LPE) have been used to prepare the materials shown, [11] although the best results obtained depend on the preparation technique. The Al-containing alloys such as AlGaAs are best prepared by LPE because of chemical problems associated with the vapor deposition of these materials and because LPE offers the possibility of using acceptor dopants, such as Ge and Si, that are not possible in VPE. On the other hand, where there is a significant lattice-parameter mismatch between the epitaxial layer and the substrate, VPE is advantageous because of the ease of compositional grading to reduce the defect density in the active region of the grown layer. For example, GaAsP is grown exclusively by VPE on GaAs or GaP substrates.

The choice of materials for laser diodes is limited by the need to incorporate heterojunctions for carrier and radiation confinement, as discussed in the previous sections. However, LED's do not require heterojunctions, although they are frequently desirable for optical communications sources (Section 6). The major problem associated with heterojunction structures arises from the interfacial lattice parameter mismatch. The simplest example is that of the joining of two hypothetical simple cubic lattice materials as shown in Fig. 4.7. A dislocation network is generated to accommodate the lattice parameter misfit, with a linear dislocation density

TABLE 4.2

BINARY AND TERNARY III-V MATERIALS
FOR 0.8-0.9 μm AND ∿1.1 μm EMISSION

λ (μm)	E_g (eV)	Material	Substrate	$\Delta a/a(\%)$ [a]
∿1.0-1.1	∿1.16	$In_{0.2}Ga_{0.8}As$	GaAs	1.38
		$In_{0.15}A_{0.85}P$	InP	0.48
		$GaAs_{0.85}Sb_{0.15}$	GaAs	1.13
∿0.88-0.91	∿1.42	GaAs	GaAs	0
∿0.82	∿1.55	$Al_{0.12}Ga_{0.88}As$	GaAs	0.017
		$GaAs_{0.86}As_{0.14}$	GaAs	0.50

[a] Lattice parameter mismatch between epitaxial layer and substrate.

$$\rho_\ell \cong \frac{\Delta a_o}{a_o} \, , \qquad\qquad (4.13)$$

where Δa_o is the small difference in lattice parameter.

The actual misfit dislocation density depends on the crystal structure, and on the elastic strain in the material; i.e., fewer dislocations are generated if the crystal remains strained to partially accommodate the lattice misfit. In addition, all of the generated dislocations do not simply lie in the interface; some are inclined and propagate into the grown layer. If the average "inclined" dislocation spacing is less than a diffusion length in the vicinity of the injected minority carriers, then the effective diffusion length, and hence the minority carrier lifetime, is degraded.

The immediate question of interest here is the effect of interfacial dislocations on the recombination velocity. Consider the (100) plane in the sphalerite structure. Assuming that each state associated with the dislocation core constitutes a nonradiative center, the calculated [12] interfacial surface recombination velocity S is

123

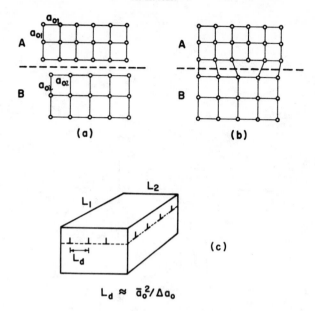

$$L_d \approx \bar{a}_o^2 / \Delta a_o$$

Figure 4.7. Schematics showing the formation of an edge misfit dislocation in joining simple cubic crystal A with lattice constant a_o and substrate B with lattice constant a_o': (a) Separate crystals; (b) formation of edge dislocation when crystals are joined; (c) formation of dislocations at edge of crystal $L_1 \times L_2$. The distance between dislocations is L_d.

$$S \cong \frac{8v_{th}\sigma_t}{a_o^2} \left(\frac{\Delta a_o}{a_o}\right) \text{ cm/s} , \qquad (4.14)$$

where v_{th} is the thermal carrier velocity and σ_t is a capture cross section. Assuming $\sigma_t = 10^{-15}$ cm^2, Eq. 4.14 predicts $S \cong 3 \times 10^5$ cm/s for $\Delta a_o/a_o = 1\%$. This value appears to be within order of magnitude agreement with the limited experimental data available. A study of InGaP/GaAs p-p heterojunctions with $\Delta a_o/a_o = 1\%$ showed that $S \sim 10^6$ cm/s, while for $\Delta a_o/a_o \lesssim 0.1\%$, $S \lesssim 10^4$ cm/s. [13] For comparison note that S for a "free" GaAs surface (i.e., unpassivated) is 10^6–10^7 cm/s.

The value of S can drastically reduce the internal quantum efficiency in the heterojunction structure if the width of the recombination region is much less than a diffusion length. The figure of merit is the parameter $SL_{n,p}/D$ (the reduced recombination velocity) for a given $d/L_{n,p}$ ratio. Consider for illustration the n-p-p structure of Fig. 4.5c. Figure 4.8 shows [14] the ratio of the radiative recombination current to the total current as a function of relevant parameters. For example, a d value of 0.3 μm is desirable for CW laser structures. A typical L_p value in moderately doped p-type GaAs is 3 μm, hence $d/L_p = 0.1$. For an internal quantum efficiency of 50%, we require $SL_p/D = 0.1$ which corresponds to $S \leq 2 \times 10^4$ cm/s. Assuming the validity of Eq. 4.14, we would require for best results that the lattice parameter mismatch not exceed about 0.1%.

The AlAs-GaAs alloy system is the only one which meets the above criterion, since the lattice parameter difference between GaAs and AlAs is $\Delta a_o/a_o = 0.14\%$, with proportionally smaller differences, of course, for reduced Al concentration differences at the heterojunctions. There are few experimental data concerning S in GaAs-AlGaAs heterojunctions; but available data do indicate that S $\approx 5 \times 10^3$ cm/s in practical laser structures, a value which is fully satisfactory for narrow recombination region devices. [15]

It is possible, however, to match lattice parameters at heterojunction structures by a judicious choice of binary, ternary, and quaternary materials. Figure 4.9 [16] shows how lattice matching combinations can be fabricated. For example, InGaAs and InGaP can be combined to provide a suitable heterojunction laser structure for $\sim$ 1.0-1.1 μm emission, as will be discussed in Section 7.

5. LASER DIODES OF $Al_xGa_{1-x}As$ FOR
ROOM TEMPERATURE CW OPERATION $(0 \leq x \leq 0.1)$

5.1 Structural Considerations

Double heterojunction laser diodes have so far routinely operated CW at room temperature with up to a 12% Al concentration in the recombination region

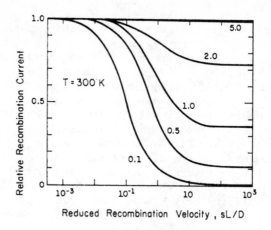

Reduced Recombination Velocity , sL/D

Figure 4.8. Ratio of the radiative current to injected electron current into a recombination region of the type shown in Fig. 4.4 as a function of the SL/D with d/L as a parameter. The p-p heterojunction barrier is assumed to be at least 0.2 eV. (Ref. 14)

(λ_L = 0.9–0.8 µm). The required threshold current density is below 4000 A/cm^2, and devices with J_{th} = 1000 - 1500 A/cm^2 are now routinely used for this purpose (value determined in broad-area lasers).

The broad area laser diode structure formed by sawing the edges perpendicular to the cleaved facets forming the mirror cavity (Fig. 4.6a) has been widely used for pulsed power operation, and in early double heterojunction laser studies,[3,17,18] but CW laser diodes now use a stripe-contact geometry exclusively for several reasons: (1) The radiation is emitted from a small region, which simplifies coupling of the radiation into fibers with low numerical aperture. (2) The operating current density can be made to be under 0.5 A because it is relatively simple to form a small active area with convenient large area contacting procedures (something more difficult to achieve with a sawed side or etched mesa). (3) The thermal dissipation of the diode is improved compared to a mesa diode because the heat-generating active region is imbedded in an inactive semiconductor medium. (4) The

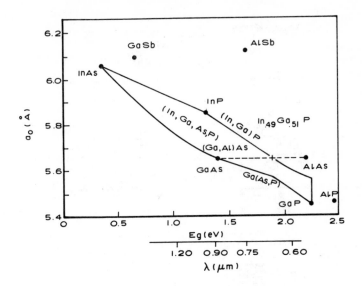

Figure 4.9. Lattice parameter and bandgap energy variation with composition for major III-V compounds. (Ref. 16)

small active diode area makes it simpler to obtain a reasonably defect-free area. (5) The active region is isolated from an open surface along its two major dimensions, a factor believed to be important for reliable long-term operation (Section 8).

The simplest structures, and those easiest to heat sink, are the planar types using one of three basic methods for defining the active area. In the first, [19] Fig. 4.10a, isolation is obtained by opening a stripe contact (typically between 10 and 20 µm wide) in a deposited SiO$_2$ film. The surface of the diode is then metallized, with the ohmic contact formed only in the open area of the surface. A second method of stripe formation [20] uses selective diffusion through an n-type region at the surface to reach to the p-type layers underneath, Fig. 4.10b. The third method [21] uses proton bombardment to form a resistive region on either side of the desired recombination region, Fig. 4.10e.

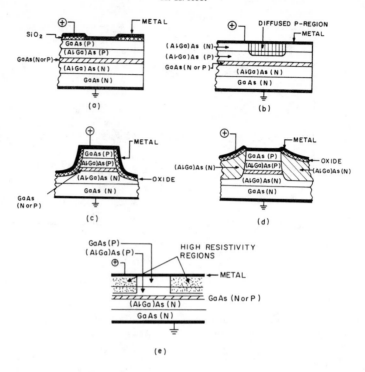

Figure 4.10. Techniques for stripe-contact diode formation.

In all of these structures, radiation and current spreading occur on either side of the stripe, depending on the resistance and thickness of the layers between the surface and the recombination region, and on the diffusion length in the recombination region. [22] Figure 4.11a shows a typical cross section of a laser diode with 10% Al in the recombination region designed for emission at ∿ 8200 Å made using oxide stripe isolation. [23] An example of the near-field radiation distribution in the plane of the junction is shown in Fig. 4.11b. While the radiation intensity is highest in the central region of the 13 μm-wide stripe, there is significant radiation extending to about 20 μm. [24]

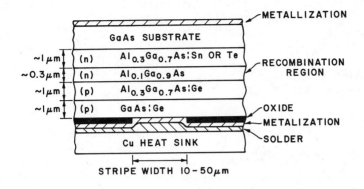

(a)

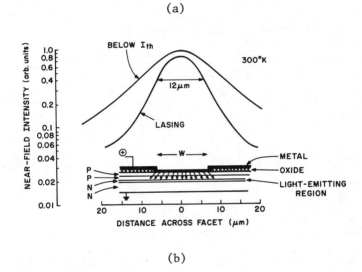

(b)

Figure 4.11. (a) Schematic laser cross section (not to scale) and (b) near-field pattern in the plane of the junction. (Ref. 24).

Diodes for CW operation are generally mounted p-side down on Cu heat sinks to maximize the thermal conductivity of the structure. A soft solder such as indium is often used to minimize strains in the devices. Diamond heat sinks can be used, but their effect in reducing the thermal resistance of the stripe-contact laser diode is relatively small.

5.2 Electrical and Optical Properties

The threshold current density of heterojunction laser diodes typically increases with temperature T as $\exp(T/T_0)$, where $T_0 \approx 50\text{–}70\,\text{K}$, while the lasing peak energy decreases with increasing temperature at a rate of $\sim 5 \times 10^{-4}$ eV K^{-1}. The maximum output from a pulsed laser occurs at a duty cycle determined by its thermal and electrical resistance, the threshold current and T_0 value. The electrical resistance of a 13 μm-wide stripe laser is generally between 0.4 and 1 Ω, and the thermal resistance is 12–20 K/W. When the power dissipation due to the diode series resistance is small compared to the power dissipation in the junction itself, the criterion for CW operation is, [25]

$$\frac{I_{th} E_g R_{th}}{e T_0} < 0.37 \quad , \tag{4.15}$$

where R_{th} is the thermal resistance.

For illustration, consider a laser diode with R_{th} = 14 K/W, E_g = 1.5 eV, and T_0 = 50 K. From Eq. (4.15), threshold currents as large as 0.88 A will provide CW operation; however, in practice, somewhat lower values than calculated are required because of series resistance losses at high current levels. For a heat sink temperature of 20°C, the upper limit for CW operation is practically about 0.7 A.

The lateral width of the emitting region can be adjusted for a desired operating level, and 100 mW of peak CW power (one facet) is obtainable for stripe widths of 100 μm. [24] However, for the typical power levels needed in optical communications (3–10 mW), stripe widths of 10–20 μm are generally used, a dimension which represents a suitable compromise between low operating currents and appropriate power emission level. In planar structures further reduction in the stripe width does not yield a corresponding threshold current reduction because of the imperfect lateral current and optical confinement.

A typical curve of power output as a function of diode current is shown in Fig. 4.12 for a diode with a 13 μm stripe. [23]   The junction temperature for such devices is only a few degrees above the heat sink temperature.   For example, a temperature differential of 7 K is calculated with a typical power input of 0.5 W at a diode current of 0.3 A, and a thermal resistance of 14 K/W.

The narrowest lateral emitting region can be obtained with a mesa configuration, particularly if AlGaAs is grown in the etched out regions to facilitate ohmic contact to the active diode area, Fig. 4.10d.   Effective stripe widths of 1-2 μm have been reported with threshold currents for CW operation of the order of 20 mW. [26]

Various techniques for modulating the optical output form laser diodes have been discussed in detail. [27]   The simplest method consists of dc biasing the diode to threshold, and modulating only the optical output above threshold. This method eliminates the delay time, t_d between current application and spontaneous light generation, given by [28]

$$t_d \stackrel{\sim}{=} \tau \, \ln \, [I/(I-I_{th})] \quad , \qquad\qquad (4.16)$$

where I and I_{th} are the amplitude of the current pulse and of the threshold current, respectively.   Modulation rates to about 1 GHz have been reported by this method. [29]

5.3   Radiation Pattern and Modal Properties

The laser diode cavity contains electromagnetic modes that are separable into two independent sets with transverse electric (TE) and transverse magnetic (TM) polarization. The modes of each set are characterized by mode numbers m, s, and q, which define the number of sinusoidal half-wave field variations along the three axes of the cavity, transverse, lateral and longitudinal, respectively.

131

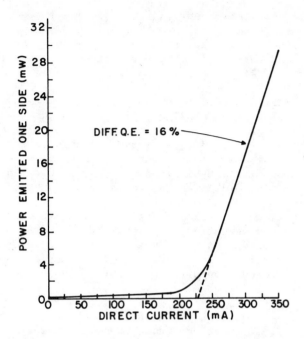

Figure 4.12. Power output as a function current for CW operation at room temperature. The diode stripe width is 13 μm. (The AlGaAs DH material is of the kind described in Ref. 23).

The allowed <u>longitudinal</u> modes (propagating between the Fabry-Perot faces in the plane of the junction) are determined from the average index of refraction, $\bar{n}$, and the dispersion seen by the propagating wave. The Fabry-Perot mode spacing for a cavity of length L and mode numbers q is:

$$\Delta\lambda/\Delta q = \frac{-\lambda_L^2}{2L(\bar{n} - \lambda \frac{dn}{d\lambda})} \quad ,$$
(4.17)

which is several Å units in typical laser diodes.

The lateral modes (in the plane of the junction) are
dependent on the method used to define the two edges of
the diode. Generally only low order modes are excited;
their mode spacings are 0.1-0.2 Å and appear as satellites
to each longitudinal mode.

The transverse modes (direction perpendicular to the
junction plane) depend on the dielectric variation perpendicular to the junction plane. In the devices discussed
here, only the fundamental mode is excited, a condition
achieved by restricting the width of the waveguiding
region (i.e., heterojunction spacing) to values well under
one micrometer. Therefore, the far-field radiation
pattern consists of a single lobe in the direction perpendicular to the junction. Higher-order transverse modes
give rise to "rabbit-ear" lobes of the type described for
wide cavity lasers which are undesirable for fiber coupling.

For a laser operating in the fundamental transverse
mode, the far-field, full angular width at half power, $\theta_\perp$,
is a function of the near-field radiation distribution.
The narrower the emitting region in the direction perpendicular to the junction plane, the larger the $\theta_\perp$. Figure 4.13
shows $\theta_\perp$ calculated as a function of d and Δn in double
heterojunction lasers. [30] In practical CW laser diodes,
$\theta_\perp$ = 35-50° with d = 0.2-0.3 μm, consistent with Δn = 0.12-0.18.
The beam width in the direction parallel to the junction
plane is typically 10° and varies only slightly with the
diode topology and internal geometry. An example of the
far-field pattern of a typical CW laser diode is shown in
Fig. 4.14a. From 1/3 to 1/2 of the power emitted from
one facet can be coupled into a step index fiber with a
numerical aperture (NA) of 0.14.

While fundamental transverse mode operation is easily
achieved, most laser diodes operate with several lateral
and longitudinal modes, as shown in Fig. 4.15 with a
spectral width of 10-15 Å, but exceptional units can emit
several milliwatts in one or two longitudinal modes. The
reason for the differences among laser diodes is believed
associated with the degree of material homogeneity. [31].

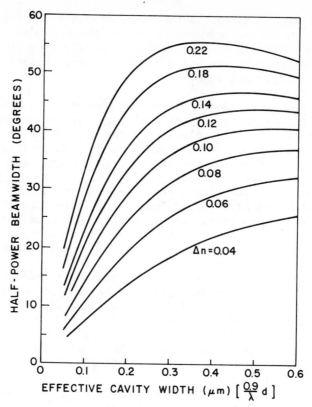

Figure 4.13. Calculated beam width in the direction perpendicular to the junction plane for double hetero-junction diodes as a function of the cavity width d and refractive index step Δn. (Ref. 30).

6. LIGHT-EMITTING DIODES

The spectral bandwidth of the LED is typically 1-2 kT (300-400 Å) at room temperature, hence at least one order of magnitude broader than the laser diode emission. Because of the increased wavelength dispersion, this limits the bandwidth for long-distance fiber communications. Furthermore, the coupling efficiency into low numerical aperture fibers is much lower than for laser diode. However, the LED has the advantage of a simpler construction and a smaller temperature dependence of the

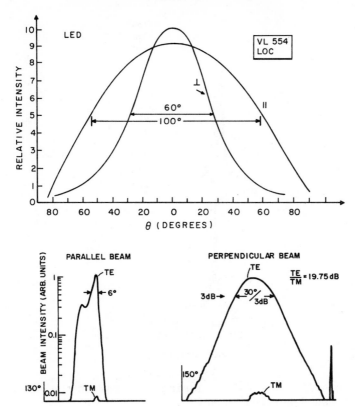

Figure 4.14. Typical radiation patterns in the plane of the junction ($\parallel$) and perpendicular to the junction ($\perp$) for (a) and edge emitting LED of the type described in Refs. 33 and 34 and (b) a laser diode of the type described in Ref. 23.

emitted power. For example, spontaneous output from a DH diode decreases by only a factor of two as the diode temperature increases from room temperature to 100°C (at constant current).

Heterojunction structures are desirable for high radiance, high speed LED's because it is possible to surround an appropriately doped recombination region

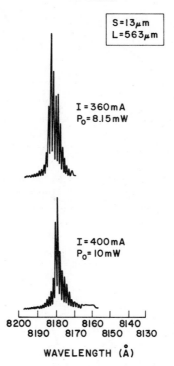

WAVELENGTH (Å)

Figure 4.15. Typical CW lasing spectrum of the AlGaAs DH lasers described in Ref. 23.

with high-bandgap-energy material having low internal absorption of the emitted radiation. Furthermore, wave-guiding leads to enhanced edge emission, resulting in a more directional beam than otherwise achieved from a surface-emitting LED.

6.1 Diode Topology

The LED topology is designed to minimize internal reabsorption of the radiation, allow high current density operation (greater than 1000 A/cm^2) and maximize the coupling efficiency into fibers. While the structures used are applicable to all materials, the bulk of the reported work has so far been on GaAs and AlGaAs devices.

Two basic diode configurations for optical communications have been reported, surface emitters [32] and edge emitters. [33,34] In the surface emitter, the recombination region is placed close to a heat sink, as shown in Fig. 4.16a, and a well is etched through the GaAs substrate to accommodate a fiber. The emission from such a diode is essentially isotropic. An improvement in the radiative efficiency of the surface emitter at the expense of fabrication complexity is obtained by providing very small etched lenses. [35]

The edge-emitting heterojunction structures uses the partial internal waveguiding of the spontaneous radiation to obtain improved directionality of the emitted power in the direction perpendicular to the junction plane. It is important, however, to restrict the active junction area to near the edge of the device in order to reduce the internal absorption of the radiation that could otherwise occur for the relatively long optical paths across a diode. Figure 4.17 shows two versions of the restricted edge-emitting diode (REED), configuration. [34] In the first, Fig. 4.17a, oxide isolation is used to form a narrow contact near one edge of the diode. While superior to the conventional stripe diode as a spontaneous radiation emitter, it can be further improved by adding a slot in the back of the emitting edge, Fig. 4.17b, which avoids a significant fraction of the radiation loss of structure (a). In all edge-emitting structures, it is also desirable to add an antireflecting film to the emitting facet to further increase the efficiency. The lateral width of the emitting region is adjusted for the fiber dimension, but is typically on the order of 50-100 μm.

Surface-emitting and edge-emitting structures provide several milliwatts of power output at current densities in excess of 1000 A/cm^2, with the current density being limited by the thermal resistance of the device (i.e., junction heating). For example, at 2000 A/cm^2, the REED diode of Fig. 17b (fabricated from an AlGaAs LOC structure with a 100 μm-wide stripe) reliably emits about 3 mW at 8000 Å. The "lensed" etched-well diode [35] has emitted 6 mW at 3400 A/cm^2, and the flat etched-well diode has emitted 2 mW at 7500 A/cm^2. [32] These values will no doubt be increased with improved technology. The coupling loss into step-index fibers with a numerical aperture of

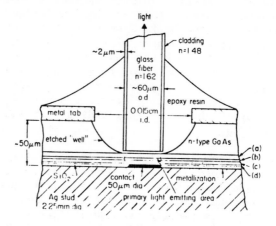

Figure 4.16. Cross-sectional drawing (not to scale) of small-area double-heterojunction electroluminescent diode coupled to optical fiber. Layer (a), n-type $Al_xGa_{1-x}As$, 10-µm thick; emitting layer (b), p-$Al_yGa_{1-y}As$, $\sim$ 1-µm thick; layer (c), p-type $Al_xGa_{1-x}As$, 1-µm thick; layer (d), p-GaAs, $\sim$ 0.5-µm thick (for contact purposes). (Ref. 32)

0.14 is about 17-20 dB for surface emitters and 14-16 dB for edge-emitting DH or LOC diodes. Since the coupling loss decreases as $(NA)^{-2}$, much more power can, of course, be coupled into larger NA fibers.

Finally, we note some noise measurements from GaAs homojunction and heterojunction LED's. Above 5 kHz, the homojunction LED was reported to be an ideal radiation source. [36] In the case of heterojunction structures, with good ohmic contacts, the measured 1/f noise would be well below the shot noise in a practical system. [37]

6.2 Modulation Capability

To first order, with the diode current constant, the dependence of the power output from an LED on modulation frequency, ω, is given by, [38]

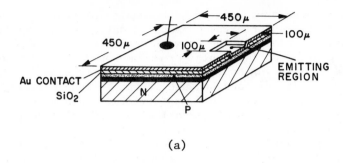

(a)

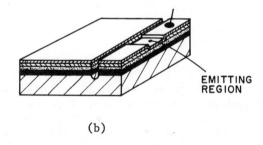

(b)

Figure 4.17. Restricted edge-emitting diodes (REED) with (a) a continuous surface and (b) a slotted back to improve the external efficiency. (Ref. 34)

$$\frac{P(\omega)}{P_o} = \frac{1}{[1 + (\omega\tau)^2]^{1/2}} \quad , \qquad (4.18)$$

where τ is the minority carrier lifetime in the recombination region, and P_o is the dc power emission value. (However, parasitic circuit elements can reduce the modulated power range below that indicated in Eq. 4.18.)

It is evident that a high speed diode requires the lowest possible value of τ, but without sacrifice in the internal quantum efficiency. As discussed in Section 2, low values of τ are obtained at high doping levels; but in

139

GaAs and related compounds, a high density of nonradiative
centers are formed when the dopant concentration approaches
the solubility limit at the growth temperature. Of the
devices reported so far, only germanium-doped DH LED's have
exhibited outstanding modulation capability (200 MHz). The
use of Ge is advantageous because it can be incorporated
into GaAs to levels of $\sim 10^{19}$ cm^{-3} (yielding minority
carrier lifetime values of 1-2 ns) without unduly reducing
the internal quantum efficiency. [39] Other dopants,
including Zn and Si, have been used to fabricate slower
diodes. [33] Figure 4.18 shows experimental and calculated
plots of $P(\omega)/P_0$ for heterojunction diodes having different
values of the minority carrier lifetime obtained by varying
the dopants in the recombination region. The data for these
broad-area diodes are in agreement with Eq. 4.18, indicating
that no other modulation-rate limiting factors were present
in those devices.

7.   LASERS AND LED'S FOR 1.0-1.1 μm EMISSION

The devices made for emission in this spectral range
are not as advanced as those using GaAs and AlGaAs because
of the difficulty in producing quality heterojunction
structures. Of the III-V materials (binary or ternary
compounds), InGaAs has received the most attention and has
been used for excellent low temperature lasers, including
CW operation at 77 K. [40] Recently vapor-grown LOC laser
structures of InGaAs/InGaP also have provided room tempera-
ture-operation by using the heterojunctions formed with the
higher-bandgap InGaP. [41] Figure 4.19a shows a cross
section of this type of laser diode, while Fig. 4.19b shows
that these lasers have room temperature $J_{th} \cong 15,000$ A/cm^2.
This is a factor of ~ 3 higher than that in GaAs DH lasers
with comparable cavity thickness of 1.3 μm but is much
lower than that of InGaAs homojunction devices. In the
lasers tested, the lattice parameter mismatch at the
heterojunctions

$$\frac{\Delta a_0}{a_0} \cong 0.2\%.$$

The effect of the interfacial surface recombination velocity
on the laser performance has not yet been established.
Lower values of J_{th} require further reductions in the

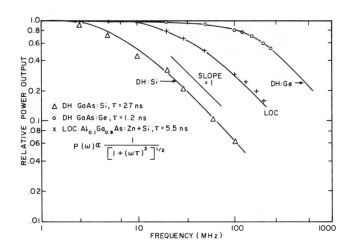

Figure 4.18. Relative power output as a function of the modulation frequency for diodes having different minority carrier lifetimes. The fast diode is of the type described in Refs. 33 and 62 , while the LOC diode is of the type described in Ref. 65. (Current current operation).

heterojunction cavity thickness and perhaps a closer lattice parameter match. These diodes could be used in the incoherent edge-emitting emission mode as well, although homojunction LED's of InGaAs have been reported with the quite satisfactory surface emission quantum efficiency of 1%. [42,43]

Other heterojunction structures studied in this spectral range are GaAsSb-AlGaAsSb LPE devices which have lased at 0.98 μm. [44] These also have been fabricated for LED operation at 1.1 μm with a quantum efficiency of about 1%. [45]

8. DIODE RELIABILITY

The two basic failure modes which may limit the operating life of electroluminescent devices are denoted catastrophic and gradual degradation. The first depends on the optical flux density and the pulse length, and is relevant only to laser diodes, while the second depends

141

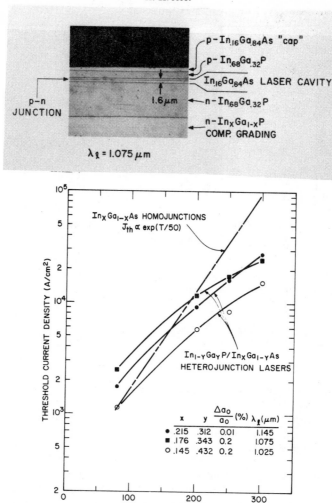

Figure 4.19. (a) Cross section of LOC InGaAs/InGaP heterojunction laser; (b) threshold current density as a function of temperature showing the reduced J_{th} compared to a homojunction. (Ref. 41).

on the current density, the duty cycle and the metallurgy
of the device. Incoherent emitters are affected by the
gradual degradation phenomena, and share with laser diodes
common properties in this regard.

8.1 Catastrophic Degradation in Laser Diodes

Facet failure due to intense optical fields is a
well known phenomenon in solid state lasers and has been
found to occur in semiconductor lasers of all types under
varying conditions. The appearance of the damaged laser
facets suggests local dissociation of the material, as
well as "cracking" in some cases. A commonly used figure
of merit is the critical damage level observed in watts
per cm of emitting facet, P_c. [46] For a given operating
condition, P_c decreases linearly with decreasing emitting
region width. For example, in double heterojunction lasers
operated with 100 ns pulses, P_c = 200 W/cm when d = 1 μm,
with P_c increasing to 400 W/cm when d = 2 μm. It is
important to note, however, that these figures represent
only average values for broad-area diodes because damage
is commonly initiated at portions of the facet that contain
mechanical flaws.

The pulse length critical damage level is also affected
by the applied pulse length, with P_c decreasing with pulse
length, t, at $t^{-1/2}$, [47] at least for pulse lengths
between 20 and 2000 ns. It is not surprising, therefore,
that catastrophic damage can occur in room-temperature
laser diodes operating CW at their maximum emission levels
(even with the relatively low power densities achievable).
Because of the non-uniform radiation distribution in the
plane of the junction in stripe-contact lasers, it is
difficult to establish simple linear power density criteria.
However, the damage threshold for 100 nsec pulse length
operation is about 10 times higher than in CW operation
of diodes selected from the same group. [24] The damage
is generally initiated in the center of the diode in the
region of highest optical intensity, as illustrated in
Fig. 4.20. Here, the damaged region for the 50 μm-wide
stripe diode is estimated to have occurred at a power level
of 2-3 mW/μm (2-4.2 x 10^5 W/cm^2 taking into account the
radiation distribution in the direction perpendicular
to the junction plane).

143

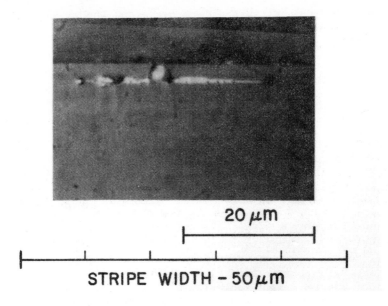

20 μm

STRIPE WIDTH - 50μm

Figure 4.20. Microphotograph showing facet damage in a laser diode operating CW at an excessive emission power level. The stripe width was 50 μm, and the damaged region is in the central portion of the stripe where the optical power density is highest. (Ref. 24)

Facet damage can be prevented by operation at reduced power levels and by the use of dielectric facet coatings. The coatings reduce the electric field at the surface by reducing the reflectivity. The experimental data [48] fit the following expression, [49]

$$\frac{P'_c}{P_c} = \frac{n(1-R')}{(1+R'^{1/2})^2} \qquad (4.19)$$

where n is the GaAs refractive index ($\sim$ 3.6), P_c is the measured critical power value for a GaAs–air interface and P'_c is the value with the facet reflectivity R'. Since J_{th} is increased by a decrease in the facet reflectivity,

R' must be chosen to give the desired facet protection without unduly degrading the device performance.

8.2 Gradual Degradation

Gradual degradation in an incoherent emitter is the reduction in the externally measured quantum efficiency. In a laser diode, the threshold current density generally increases, and the differential quantum efficiency decreases, resulting in a reduced power output for a given current density without evidence of facet damage. Of course, if a laser is operated near threshold, it may cease lasing altogether for a small increase in J_{th}; however, in this case a small increase in current can usually recover the initial power value. Therefore, the definition of degradation in a laser diode is somewhat arbitrary, since current adjustments can keep a laser operating despite changes in its efficiency with time.

Gradual degradation can occur in both homojunction and heterojunction lasers. [50] The available evidence indicates that the internal quantum efficiency is reduced by the formation of nonradiative centers within the recombination region [51] and that the internal absorption coefficient can increase. It is well established that the minority carrier lifetime decreases as degradation progresses, suggestive of a decrease in τ_{nr}. This is consistent with the observed reduction in the external quantum efficiency of diodes operating in the spontaneous emission mode. Whether lasing or not, diodes of a given construction degrade similarly at a given current density. It is known that the presence of a p-n junction is not needed for the occurrence of gradual degradation, since optically excited GaAs shows evidence of nonradiative center formation in the region where electron-hole pair recombination occurs. [52]

An important characteristic of the degradation process is that it is spatially nonuniform, and that it is highly dependent on the type of semiconductor, the degree of perfection of the material, and the method used to fabricate and assemble the diode. For example, dislocations [53] and exposed active region edges should be minimized, [23] and strains introduced in the process of assembling diodes

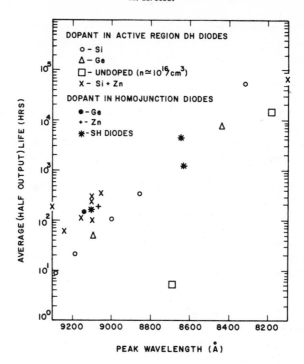

Figure 4.21. Average (half-output) life of incoherent emitters stressed at $\sim 1000 A/cm^2$ as a function of emission wavelength. For each doping type the shift to shorter wavelength represents additions of Al to the recombination region. (Ref. 54)

with "hard" solders should be avoided. The use of $Al_xGa_{1-x}As$ ($x \approx 0.1$) is found to be very desirable to improve diode life as shown in Fig. 4.21. [54] Work has also been reported on the reduction of strain due to the difference in the thermal expansion of GaAs and AlGaAs by the addition of P to the AlGaAs regions surrounding the GaAs recombination region. [55] This strain may contribute to degradation as do external constraints. [56]

All of the available evidence suggest that the gradual degradation process is initiated by flaws, initially present in the recombination region of the diode, which grow in size, and/or by defects which indiffuse from external regions adjacent to the active region. One prominent effect in some degraded lasers is the formation of "dark lines" in which the luminescence is greatly reduced. [57] These regions have been identified as large dislocation networks which, starting at existing smaller dislocation networks, grow by the immigration of vacancies or interstitials. [58] In addition, more dispersed nonradiative centers such as native point defects could contribute to the degradation process.

The origin of these point defects is still the subject of speculation. Gold and Weisberg, [59] in their GaAs tunnel diode degradation studies, suggested that nonradiative electron-hole recombination at an impurity center could result in its displacement into an interstitial position, leaving a vacancy behind. This is the "phonon-kick" model in which multi-phonon emission gives an intense vibration of the recombination center which effectively reduces its displacement energy. Applying this mechanism to the electroluminescent efficiency degradation, the assumption was that the vacancy and interstitial formed would have a large cross section for nonradiative recombination. Repeated nonradiative recombination would gradually move it to internal sinks.

Evidence is accumulating that the energy released in the electron-hole recombination process is in fact a factor in the degradation process.

(1) The InGaAs LED work showed that despite a high dislocation density, the degradation rate decreased with decreasing bandgap energy (a favorable development for 1.1 μm emitters). [60]

(2) Quantitative support for this hypothesis has also been obtained by showing that the damage introduced into GaAs diodes by irradiation with 1 MeV electrons would anneal much more readily under forward bias. The activation energy formation of the unidentified defects was reduced to 0.34 eV from 1.4 eV in the presence of electron-hole recombination in theforward-biased diode. [61]

The following is a working hypothesis for explaining the major degradation effects. In essence, the "phonon-kick" process uses part of the energy released in non-

radiative recombination to enhance the diffusion of
vacancies, or more probably interstitials. (Whether any
point defects are actually formed within the recombination
region remains unclear.) Hence, if nonradiative electron-
hole recombination occurs, for example, at the damaged
surface of a diode (as in the case of the sawed-edge
diode experiment described in [23]), it accelerates the
motion of point defects into the active region of the
device. Thus, the stoichiometry of the material in or in
close proximity to the active region can affect the local
density of defects, [62] if the point defect concentration
is substantially above the equilibrium value at room
temperature. Differences in initial stoichiometry may
contribute to the improved degradation resistance of
AlGaAs diodes. Finally, regions where nonradiative recombination occurs will tend to grow in size, leading to the
strongly non-uniform degradation process commonly observed.

From the available results, control of gradual degradation appears to be primarily a metallurgical problem
with most of the contributing factors known.

The progress made to date suggests that the operating
lifetimes for both LED's and CW laser diodes are coming
in the range of practical fiber systems requirements.
AlGaAs LED's with half-lives in excess of 10,000 hrs are a
reality, and laboratory samples of CW laser diodes have
operated for periods of time in excess of 10,000 hrs,
although current increases are generally needed to maintain
a power output of a few milliwatts. Facet erosion has been
suggested as a factor in the degradation seen over many
thousands of hours. [23,24] In fact, recent AlGaAs
CW lasers with dielectric facet coatings (to eliminate facet
damage) operated with no degradation for thousands of
hours. [24] The major problem with regard to systems
application is the reliable forecasting of diode life based
on accelerated testing. Some data show that the diode
degradation rate increases with operating temperature, [63]
which would provide one method for accelerating life tests
if in fact the degradation rate varies consistently with
temperature.

REFERENCES

1. H. Kressel and H. Nelson, RCA Review 30, 106 (1969).

2. I. Hayashi, M. B. Panish, and P. W. Foy, IEEE J.
   Quantum Electron. 5, 211 (1969).

3. Zh. I. Alferov, V. M. Andreev, E. L. Portnoi, and
   M. K. Trukan, Sov. Phys. Semiconductors 3, 1328
   (1969). [Trans.: Sov. Phys. Semiconductors 3, 1107
   (1970].

4. (a) M. H. Pilkuhn, Physica. Status Solidi 25, 9 (1969);
   (b) A. Yariv, Quantum Electronics, John Wiley and
       Sons, Inc., New York, 1967;
   (c) C. H. Gooch (Ed.), GaAs Lasers, John Wiley and
       Sons, Inc., New York, 1969;
   (d) H. Kressel, a chapter in Advances in Lasers,
       Vol. 3, A. K. Levine and A. J. DeMaria, Eds.,
       Marcel Dekker, New York, 1971;
   (e) J. I. Pankove, Optical Processes in Semiconductors,
       Prentice-Hall, New York, 1971;
   (f) H. Kressel, in Laser Handbook, F. T. Arecchi and
       E. O. Schulz-DuBois, Eds., North Holland,
       Amsterdam, 1972;
   (g) P. G. Eliseev, Sov. J. Quantum Electron. 2,
       505 (1973);
   (h) M. B. Panish and I. Hayashi, Applied Solid State
       Science, R. Wolfe, Vol. 4, Academic Press,
       New York, 1974.

5. Y. P. Varshni, Phys. Status Solidi 19, 353 (1964).

6. H. Kressel, H. F. Lockwood, F. H. Nicoll, and
   M. Ettenberg, IEEE J. Quantum Electron. QE-9, 383 (1973).

7. H. Kressel, J. K. Butler, F. Z. Hawrylo, H. F. Lockwood,
   and M. Ettenberg, RCA Review 32, 393 (1971).

8. H. F. Lockwood, H. Kressel, H. S. Sommers, Jr., and
   F. Z. Hawrylo, Appl. Phys. Letters 17, 499 (1970).

9. G. H. B. Thompson and P. A. Kirkby, IEEE J. Quantum
   Electron. QE-9, 311 (1973).

10. J. J. Tietjen, Annual Review of Materials Science 3, 317 (1973). Published by Annual Reviews, Inc., Palo Alto, Calif.

11. A comprehensive review is presented by H. Kressel and H. Nelson, "Properties and Applications of III-V Compound Films Deposited by Liquid Phase Epitaxy," in Physics of Thin Films, Vol. 7, G. Hass, M. Francombe, and R. W. Hoffman, Eds., (Academic Press, New York, 1973).

12. D. B. Holt, J. Phys. Chem. Solids 27, 1053 (1966).

13. R. U. Martinelli and H. G. Olsen, unpublished.

14. R. D. Burnham, P. D. Dapkus, N. Holonyak, Jr., D. L. Keune and H. R. Zwicker, Solid-State Electron. 13, 199 (1970).

15. M. Ettenberg and H. Kressel, to be published.

16. R. U. Martinelli and D. G. Fisher, Proc. IEEE 62, 1339 (1974).

17. I. Hayashi, M. B. Panish, P. W. Foy, and S. Sumski, Appl. Phys. Letters 17, 109 (1970).

18. H. Kressel and F. Z. Hawrylo, Appl. Phys. Letters 17, 169 (1970).

19. J. C. Dyment, Appl. Phys. Letters 10, 84 (1967).

20. H. Yonezu, I. Sakuma, K. Kobayashi, T. Kamejima, M. Urno, and Y. Nannichi, Japan. J. Appl. Phys. 12, 5185 (1973).

21. J. C. Dyment, L. A. D'Asaro, J. C. North, B. I. Miller, and J. E. Ripper, Proc. IEEE (Letters) 60, 726 (1972).

22. B. W. Hakki, J. Appl. Phys. 44, 5021 (1973).

23. I. Ladany and H. Kressel, Appl. Phys. Letters 25, 708, (1974).

24. H. Kressel and I. Ladany, RCA Review 36, 230 (1975).

25. R. W. Keyes, IBM Journal Res. Dev. 15, 401 (1971).

26. T. Tsukada, J. Appl. Phys. 45, 4899 (1974).

27. T. L. Paoli and J. E. Ripper, Proc. IEEE 58, 1457 (1970).

28. K. Konnerth and C. Lanza, Appl. Phys. Letters 4, 120 (1964); J. E. Ripper, J. Appl. Phys. 43, 1762 (1972).

29. M. Chown, A. R. Goodwin, D. F. Lovelace, G. B. H. Thompson, and P. R. Selway, Electronics Letters 9, (25 January) 1973.

30. J. K. Butler, unpublished.

31. See, for example, P. G. Eliseev and V. P. Strakhov, JETP Letters 16, 428 (1972).

32. C. A. Burrus and B. I. Miller, Opt. Commun. 4, 307 (1971).

33. M. Ettenberg, H. F. Lockwood, J. Wittke, and H. Kresse, 1973 International Electron Device Meeting, Washington, D.C., Technical Digest, p. 317.

34. H. Kressel and M. Ettenberg, Proc. IEEE (to be published, 1975).

35. F. D. King and A. J. Springthorpe, J. Electron. Materials 4, 243 (1975).

36. J. Conti and M. J. O. Strutt, IEEE J. Quantum Electron. QE-8, 815 (1972).

37. T. P. Lee and C. A. Burrus, IEEE J. Quantum Electron. QE-8, 370 (1972).

38. Y. S. Liu and D. A. Smith, Proc. IEEE 63, 542 (1975). J. Wittke independently derived this expression for heterojunction structures. We are indebted to Dr. Wittke for the P(ω) measurements given here.

39. H. Kressel and M. Ettenberg, Appl. Phys. Letters 23, 511 (1973).

40. C. J. Nuese, M. Ettenberg, R. E. Enstrom, and H. Kressel, Appl. Phys. Letters 24, 224 (1974).

41. C. J. Nuese and G. H. Olsen, Appl. Phys. Letters 26, 528 (1975).

42. C. J. Nuese and R. E. Enstrom, IEEE Trans. Electron. Devices ED-19, 1067 (1972).

43. R. E. Nahory, M. A. Pollack, and J. C. DeWinter, J. Appl. Phys. 46, 775 (1975).

44. K. Sugiyama and H. Saito, Japan. J. Appl. Phys. 11, 1057 (1972).

45. R. E. Nahory, M. A. Pollack, E. D. Beebe, and I. C. DeWinter, to be published.

46. H. Kressel and H. P. Mierop, J. Appl. Phys. 38, 5419 (1967).

47. Extensive data are given by P. G. Eliseev in Semiconductor Light Emitters and Detectors, A. Frova, Ed., North-Holland Publishing Co., Amsterdam, 1973.

48. M. Ettenberg, H. S. Sommers, Jr., H. Kressel, and H. F. Lockwood, Appl. Phys. Letters 18, 571 (1971).

49. B. W. Hakki and R. Nash, J. Appl. Phys. 45, 3907 (1974).

50. A comprehensive discussion of the literature until 1973 was presented by H. Kressel and H. F. Lockwood, J. de Physique, C3, Suppl., 35, 223 (1974).

51. H. Kressel and N. E. Byer, Proc. IEEE 38, 25 (1969).

52. W. D. Johnston and B. I. Miller, Appl. Phys. Letters 23, 192 (1973).

53. H. Kressel, N. E. Byer, H. F. Lockwood, F. Z. Hawrylo, H. Nelson, M. S. Abrahams, and S. H. McFarlane, Met. Trans. 1, 635 (1970).

54. M. Ettenberg, H. Kressel, and H. F. Lockwood, Appl. Phys. Letters 25, 82 (1974).

55. J. C. Dyment, F. R. Nash, C. J. Hwang, G. A. Rozgonyi, R. L. Hartman, H. M. Marcos, and S. E. Haszko, Appl. Phys. Letters 24, 481 (1974).

56. R. L. Hartman and A. R. Hartman, Appl. Phys. Letters 23, 147 (1973).

57. B. C. DeLoach, Jr., B. W. Hakki, R. L. Hartman, and L. A. D'Asaro, Proc. IEEE 61, 1042 (1973).

58. P. Petroff and R. L. Hartman, Appl. Phys. Letters 23, 469 (1973).

59. R. D. Gold and L. R. Weisberg, Solid-State Electron. 7, 811 (1964).

60. M. Ettenberg and C. J. Nuese, J. Appl. Phys. 46, 2137 (1975).

61. D. V. Lang and L. C. Kimerling, Phys. Rev. Letters 33, 489 (1974).

62. M. Ettenberg and H. Kressel, Appl. Phys. Letters 26, 478 (1975).

63. R. L. Hartman and R. W. Dixon, Appl. Phys. Letters 26, 239 (1975) and references therein.

64. C. J. Nuese, H. Kressel, and I. Ladany, IEEE Spectrum 9, 28 (1972).

65. H. Kressel, H. F. Lockwood, and F. Z. Hawrylo, J. Appl. Phys. 43, 561 (1972).

Chapter 5 - Photodetectors for Fiber Systems

S. D. Personick

Bell Telephone Laboratories
Holmdel, NJ   07733

1. Introduction

In this chapter we shall be discussing photodetection,-
the conversion of optical power into electrical current.
Since we are concerned with the use of photodetectors in
fiber systems, we shall restrict our discussion to devices
which appear at this time to be most suitable in that
application. We shall develop the principles of photodetection from three points of view. The devices themselves will be discussed with regard to the physical
principles which govern their operation,-including some
of the trade-offs which have led to the physical
structures which are currently used. Next we shall
develop a circuit model for photodetectors. This model
will allow us to design the electronic circuitry to
amplify and process the detector output in an efficient
manner. Finally, we shall develop a statistical model
which describes the "noisy" relationship between the
incident light variations and the emitted current variations.
In a given circuit application, we will be able to predict
the minimum average light level required for satisfactory
information extraction.

2. The Photoelectric Effect

We shall begin our discussion of photodetection with
the familiar photoelectric effect. Consider a photoemissive
material, which by definition will emit electrons when
illuminated by light having a sufficiently short wavelength.
Such a photoemissive material can be fabricated into a

155

photocathode as part of a vacuum photodiode as shown in
Figure 5.1. The fact that electrons are emitted only
when the light is of sufficiently short wavelength,
regardless of intensity (watts/cm^2), can be explained in
terms of hypothetical packets of light energy called photons.

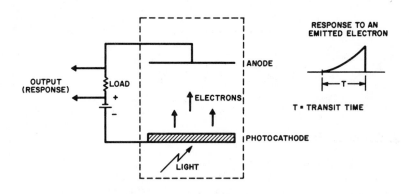

FIG. 5.1 VACUUM PHOTODIODE

We begin by proposing that electrons in the photocathode
absorb energy from the incident light in fixed packets
having magnitude hν Joules, where h = Plancks constant
and where ν is the optical frequency which is equal
numerically to the speed of light in free space divided
the free space wavelength of the light. For example, at a
wavelength of 1 micron, the photon energy is approximately
2x10^{-19} Joules or about 1.25 electron volts. When the
electron absorbs energy, it can escape from the photocathode
surface, provided this absorbed energy is large enough to
overcome the forces which bind the electron to the photocathode. The minimum required energy is called the work
function. The actual required energy is given by the work
function plus any additional energy required for the electron
to reach the photocathode surface or which will remain with the
electron as kinetic energy after escaping. If we equate the
minimum required energy (work function) to hν , we observe
indeed that there is a minimum frequency (maximum wavelength)
for photoemission. In the vacuum photodiode, emitted electrons
are collected at the anode. During the transit from cathode
to anode, a displacement current flows producing a voltage

which is developed across the load resistor. The speed of
the vacuum photodiode is determined by the transit time.
The sensitivity of the photodiode is specified by one of
two related quantities. The quantum efficiency is the fraction
of incident photons which liberate electrons. The
responsivity is the average emitted current divided by the
average incident power. It is equal to the (electron charge/
photon energy) x quantum efficiency (coulombs/joule= amps/
watt). Curves of quantum efficiency and responsivity for
some typical photocathode surfaces are shown in Figure 5.2.

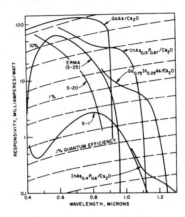

**FIG. 5.2: RESPONSIVITY AND QUANTUM EFFICIENCY
OF TYPICAL PHOTOCATHODES**

3. Photodiodes

Vacuum photodiodes are useful in application where
size is not critical. In fiber systems, microscopic
semiconductor photodiodes appear to be more appropriate
and economical. The basic photodiode is essentially a
backbiased PN junction as shown in Figure 5.3. Important

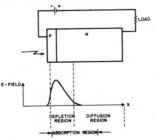

FIG. 5.3: PN DIODE

157

features of such a junction are the depletion region containing a relatively high field and the absorption region
where incident light is absorbed (photons are captured).
The depletion region is formed by immobile positively
charged donor atoms on the n side of the junction and
immobile negatively charges acceptor atoms on the p side.
The width of the depletion region depends upon the doping
concentrations. The lower the doping, the wider the
depletion layer. The absorbtion region position and width
depends upon the wavelength of the incident light and on
the material out of which the diode is made. The more
strongly light is absorbed, the shorter the absorbtion
region. The absorbtion region may extend completely through
the diode if light is only weakly absorbed. When photons
are absorbed, electrons are excited from the valence band
into the conduction band. Thus an electron hole pair is
created. If the pair is created in the depletion region,
as shown in Figure 5.4, then these carriers will separate

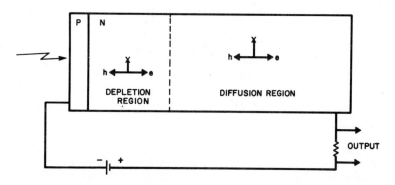

X = HOLE-ELECTRON PAIR CREATED

FIG. 5.4: CARRIER GENERATION

(drift) under the influence of the field in the depletion
region. This will produce a displacement current in the
biasing circuit. If a hole-electron pair is created outside
of the depletion region, then the hole will diffuse toward
the depletion region and then be collected. Since diffusion
is very slow compared to drift, it is desirable that most of
the light be absorbed in the depletion region. Thus, the
diode designer tries to make the depletion region long by

decreasing the doping in the n layer. This leads to such
a lightly doped n layer that it can be considered intrinsic.
Therefore, we end up with a PI junction. If for the practical
purpose of making a low resistance ohmic contact, we add on
a highly N doped region as shown in Figure 5.5,-we end up with

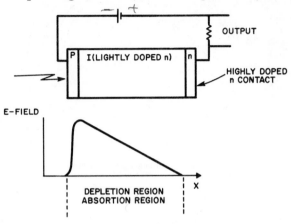

FIG. 5.5: PIN DIODE

the familiar PIN structure. It should be pointed out that
any light which is absorbed in this added n region will not
result in delayed carriers diffusing into the depletion region
since in the highly doped n region, carrier lifetime is short.
That is carriers produced there are likely to combine with
electrons before they reach the depletion region.

In the PIN diode there is often a trade-off between
quantum efficiency (fraction of light absorbed) and speed
of response. For high quantum efficiency, we desire a long
absorbtion region. For fast speed we require short drift
times and thus a short absorbtion region with high carrier
velocities. Figure 5.6 shows the quantum efficiency and
responsivity of some typical high speed photodiodes.

4. Avalanche Photodiodes

If one considers an ideal photodiode in which every
incident photon produces a photoelectron, then at the
wavelength 1 μm the responsivity is about .8 ampere per
watt. As we shall see later, the most sensitive receivers
operate at optical input levels of about a nanowatt. Thus,

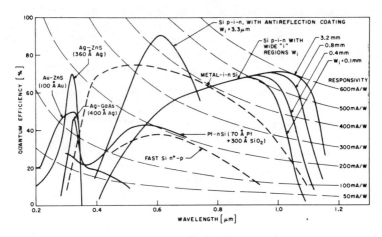

FIG. 5.6: QUANTUM EFFICIENCY OF TYPICAL
HIGH SPEED PHOTODIODES

the photoelectron current is a few nanoamperes or less.
Such small currents are very difficult to process electroni-
cally without adding excessive amplifier noise. Thus, it
is desirable to use some mechanism to increase the photodiode output current before amplification. Such a mechanism
is avalanche gain.

Consider the avalanche photodiode shown in Figure 5.7.
At very high reverse bias voltages, carriers traveling through

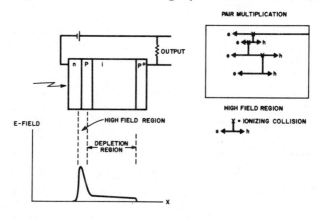

FIG. 5.7: AVALANCHE PHOTODIODE

the high field portion of the depletion region can produce
new carriers by the mechanism of collision ionization. This
is the phenomenon that leads to avalanche breakdown in ordinary
reverse biased diodes. These new carriers can themselves
produce additional new carriers by the same collision ionization mechanism. Thus, an initial, primary, hole electron
pair may result in tens, hundreds or even more secondary
pairs. The result is an effective amplification of the
photodiode output current. This amplification is very useful,
but has some shortcomings when compared to what might be
thought of as an ideal electron multiplying mechanism. Such
a mechanism would replace each hole electron pair by exactly
G secondary hole-electron pairs. That is, the pair multiplication would be deterministic. In an actual avalanche
photodiode, the multiplication is random. That is, each
primary pair gets replaced by a random number of secondary
pairs whose average number might be G (same as the ideal
multiplier) but whose variance is large. This randomness,
or unpredictability of the amplified current is a degradation
which limits the sensitivity of an optical receiver below
what it could be for an ideal multiplication device. In
designing the photodiode, one wishes to minimize the randomness.
First one selects very uniform materials so that the multiplication is not a random function of the path taken by the
carriers as they drift through the high field region. Also,
since the multiplication statistics are a function of where
the primary hole-electron pair is created relative to the
high field region, one designs the device so that most primary
photo pairs are created outside the high field region on the
side where the more strongly ionizing carrier (hole or
electron) will drift into the high field region. These
precautions result in a diode with good multiplication
statistics (as we shall soon see more quantitatively).

In addition to good multiplication statistics, speed
of response is important in avalanche photodiodes. Like PIN
diodes, initial (primary) hole electron pairs should be
generated mostly in a region with a sufficiently high drift
field to bring the carriers to the high field (multiplication)
region as quickly as possible. Carriers which are generated
in very low field regions result in so called diffusion tails
in the diode response. From the diagram (Figure 5.7) we see
that carriers move back and forth across the high field
region many times as the multiplication process proceeds.
In practical diodes available today, the devices exhibit a
trade-off between gain and bandwidth (response speed). A

gain bandwidth product can be approximately ascribed to these devices which is on the order of 100GHz.

The structure of avalanche diodes usually includes a "guard ring". The purpose of this element is to prevent low breakdown voltages and excessive leakage at the junction edges,-by lowering the fields in those regions.

Most avalanche detectors available today are made of silicon. The best of these devices provide good multiplication statistics (for optical fiber receiver applications) and gains of a few hundred (enough for most fiber communication applications). In addition, with proper anti-reflection coatings, these detectors provide quantum efficiencies approaching 100%.

5. Photodiode Circuit Model

In a typical optical fiber system application, a photodiode would be connected as shown in Figure 5.8. The power supply, bypassed by an appropriate capacitor is placed in series

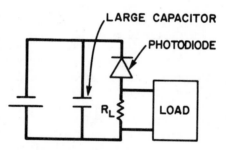

FIG. 5.8: PHOTODIODE CIRCUIT

with the diode and a load circuit. The load circuit typically consists of a large load resistor and an ac coupled amplifier in parallel. The bias voltage is adjusted for a backbiased condition on the photodiode. Since under typical conditions, the dc signal is a few microamperes or less, most of the bias voltage appears across the diode. Depending upon the diode structure and the optical wavelength, the responsivity

may or may not be a strong function of the applied bias.
Typically the bias adjustment is not critical over a reasonable range (say 10% above or below the rated bias condition).
When light is incident upon the diode, current flows in
proportion to the responsivity.

As carriers are generated and separate in the device,
they create a voltage across the diode junction capacitance,
which is discharged through the load circuitry. An ac
equivalent circuit for fiber applications is shown in
Figure 5.9. We see that the diode is essentially a current

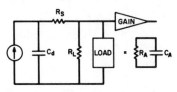

C_d = DIODE JUNCTION CAPACITANCE
R_S = DIODE SERIES RESISTANCE
R_L = PHYSICAL LOAD RESISTOR
R_A = AMPLIFIER INPUT RESISTANCE
C_A = AMPLIFIER SHUNT CAPACITANCE
C_T = $C_A + C_d$

$R = 1/(\frac{1}{R_L} + \frac{1}{R_A})$

FIG. 5.9: EQUIVALENT CIRCUIT

source in parallel with its junction capacitance (typically
a few pF). A small series resistance is shown (a few ohms)
which is negligible in most applications. There is also
some shunt resistance which typically is large enough to
be neglected except at very low frequencies. We have not
yet mentioned the load circuitry, i.e., the amplifier to
be used with the photodiode. This is a subject for the
next lecture. The optimal design of an amplifier is based
on a number of practical considerations, as well as the major
consideration of amplifier noise. For the moment, let us
assume the following suboptimal (from a noise point of view)
amplifier shown in Figure 5.9. It is modeled as a resistor
in parallel with a capacitor followed by an ideal, infinite
impedance amplifier. If the amplifier has gain $A(f)$, where
f is a baseband frequency, then the output voltage at
frequency f in response to a component in the detector current
at frequency f is given by the simple formula

163

$$V(f) = I(f)Z(f)A(f) \qquad (5.1)$$

where $Z(f)$ is the impedance of the load at frequency f

$$Z(f) = \frac{1}{1/R + j2\pi f C_{total}} \qquad (5.2)$$

Furthermore, if the optical power falling on the detector is $p(t)$ watts and if the detector responsivity is R (amp/watt) then the amplifier output voltage is

$$v(t) = p(t) R * h_{diode}(t) * h_{amp\text{-}load}(t)$$

where

* indicates the convolution of two impulse responses, $h_{diode}(t)$ is the detector impulse response (Fourier transform of the detector frequency response)

and

$h_{amp\text{-}load}(t)$ is the load-amplifier impulse response (Fourier transform of $Z(f)A(f)$)

It is important to recognize that $p(t)$, the received optical power varies in time at a baseband rate (according to the modulation) not at an optical rate. If $p(t)$ varies slowly, then $v(t)$ is just a scaled replica of $p(t)$.

For example, consider Figure 5.10. We assume that $p(t)$ varies sinousoidally about some mean value with peak amplitude

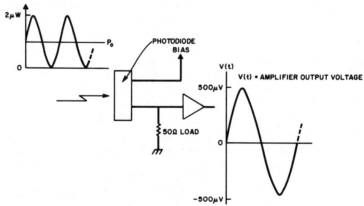

FIG. 5.10: EXAMPLE: SINOUSOIDAL MODULATION

(of the sinousoidal part) of $1\mu W$, and at the frequency 1MHz. We assume a detector responsivity of 1 amp/watt. We further assume that the detector frequency response is unity to frequencies well beyond 1MHz (no detector roll-off). The amplifier input impedance is dominated by the 50 ohm resistor at these frequencies. The amplifier gain is 20dB. Thus, the amplifier output voltage is a sinousoid with a peak value of $500\mu V$. Clearly, further amplification is needed to process this signal.

As a final comment in this section,-the above circuit model assumes that the detector output current is a linearly filtered version of the received optical power. For PIN detectors, this is an excellent approximation, violated only at optical power levels which are too high for consideration in fiber communication applications.

6. Avalanche Photodiode Circuit Model

Much of the preceding discussion of the PIN diode circuit model carries over to avalanche diodes. Figure 5.11 shows an avalanche diode biasing circuit in schematic form.

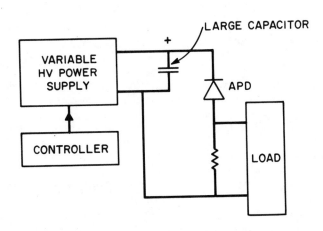

FIG. 5.11: AVALANCHE PHOTODIODE CIRCUIT

The important distinction between this and the PIN circuit is the very strong dependence of the device responsivity on the bias voltage. A typical variation of avalanche detector responsivity with bias voltage is shown in Figure 5.12. That is the reason for the temperature compensating power

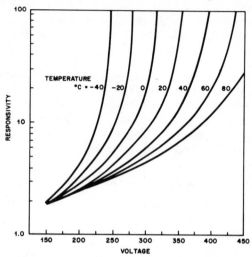

FIG. 5.12: RESPONSIVITY VS VOLTAGE FOR TYPICAL APD

supply. As the temperature varies, the detector breakdown voltage varies, and therefore the proper bias required to maintain a fixed avalanche gain varies. It should also be noted that this proper bias may be hundreds of volts or more for some avalanche detectors. The design of the voltage control circuit depends upon the application. In some cases, the average current flowing in the bias circuit can be monitored and fed back to the controller. Sometimes, an auxilliary avalanche diode, matched to the active diode, is monitored, to determine its avalanche gain. If the matching is good, then the bias voltage, which is applied to both diodes, can be varied so as to produce the proper gain in the auxilliary diode and therefore, (hopefully) in the active diode. The need to provide a controlled high voltage bias is a serious drawback of avalanche photodiodes. However, this is sometimes offset by the enhanced responsivity,-due to the carrier multiplication.

An equivalent circuit for an avalanche photodiode is given in Figure 5.13. It is identical to the PIN diode circuit with the following caution. Since the avalanche

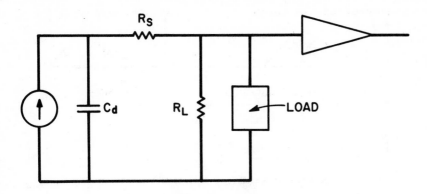

FIG. 5.13: APD EQUIVALENT CIRCUIT

multiplication is bias sensitive, nonlinearity (saturation) can occur if the applied optical signal variations are very large. Usually avalanche detectors are used to achieve the ultimate in sensitivity,-implying small optical signals. However, where dynamic range requirements are excessive, caution must be exercised.

7. <u>Noise in Photodiodes</u> (without avalanche gain)

In the previous sections, we discussed the concept of responsivity. When a given power p(t) (watts) was incident upon the photodiode, a certain average current i(t) was generated by the diode. This current was proportional to the power,-the responsivity (amps/watt) being the proportionality constant.

On the other hand, we know that the diode current consists of the sum of the displacement currents of individual hole electron pairs generated within the device (see Figure 5.14). The times at which these hole electron pairs are generated are not precisely predictable. Thus, although the <u>average</u> current pulse response (defined by adding up a large number of current pulse responses and dividing by the number which we added) looks like the input power pulse, any individual pulse response differs from the average by some unpredictable amount. We can call the difference between the actual pulse in any given response, and the average pulse response,-a

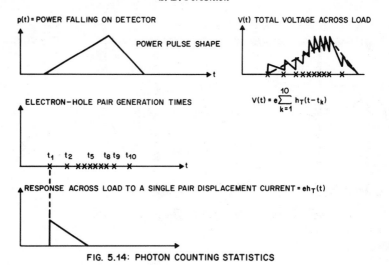

p(t) = POWER FALLING ON DETECTOR

POWER PULSE SHAPE

V(t) TOTAL VOLTAGE ACROSS LOAD

ELECTRON-HOLE PAIR GENERATION TIMES

$$V(t) = e \sum_{k=1}^{10} h_T(t-t_k)$$

t_1 t_2 t_5 t_8 t_9 t_{10}

RESPONSE ACROSS LOAD TO A SINGLE PAIR DISPLACEMENT CURRENT = $eh_T(t)$

FIG. 5.14: PHOTON COUNTING STATISTICS

signal dependent noise e(t). This noise is signal dependent because its statistics are a function of the average pulse. In order to design receivers which perform reliably, we must know something about the statistics of this signal dependent noise. In the next section, we shall formulate a noise model for PIN and avalanche photodiodes.

8. Mathematical Noise Model

We begin our model by assuming a fixed pulse of optical power p(t) watts is incident upon the photodiode. In response to this power pulse, pairs of holes and electrons are generated in the diode at times $\{ t_n \}$.

These hole electron pairs produce displacement currents which build up a voltage across the diode load. For simplicity, we shall assume that the displacement current produced by a hole electron pair is the same as that produced by any other hole electron pair. Referring to Figure 5.14, the voltage developed across the load is given by the following formula

$$v_{load}(t) = e \sum_{n=1}^{N} h_T\{t-t_n\} \qquad (5.3)$$

In equation (5.3), $eh_T(t)$ is the overall impulse response of the diode and the load, to an electron hole pair generated at time t=0. The total number of hole-electron pairs generated is N. Both N and the set of generation times $\{t_n\}$ are random quantities. It has been shown that these random generation times form what is referred to by statisticians as a Poisson random process, with a time varying rate. Basically, this means the following. If we divide the time axis into small intervals of length dt, as shown in Figure 5.15, then in any interval either 1 or 0 hole electron pairs will be created within the diode. The

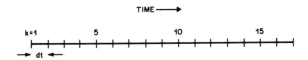

TIME ⟶

PROB "COUNT" IN INTERNAL $dt \cong \lambda(kdt)\,dt$

"NO COUNT" $\cong 1 - \lambda(kdt)\,dt$

$\lambda(t)$ = INTENSITY OF THE PROCESS

FIG. 5.15: POISSON RANDOM PROCESS

probability that one pair is created, is given by $\lambda(t)dt$ (where $\lambda(t) = (\eta/h\nu)\,p(t)$ = average # electrons generated per second). The probability, that no pair is created, is given by $1 - \lambda(t)dt$. These formulas assume that the intervals are so small that the probability that more than one pair is created in an interval is negligible. In addition, in the statistical model, it is assumed that whether or not a given interval gives rise to a hole electron pair is independent of whether or not any other interval gives rise to a pair.

From these few assumptions, a number of important conclusions follow. First, the total number of hole-electron pairs created during the interval of time $[t, t+T]$ is a random variable having the following probability distribution:

$$\text{Prob } (N=n) = \frac{\Lambda^n e^{-\Lambda}}{n!} \qquad (5.4)$$

where
$$\Lambda = \int_t^{t+T} \lambda(t)dt = \int_t^{t+T} \frac{\eta}{h\nu} p(t)dt$$

This is the familiar Poisson distribution with mean value of n equal to Λ.

Furthermore, referring to equation (5.3) the <u>average</u> voltage across the load can be derived as

$$V_{\text{load average}} = \frac{\eta e}{h\nu} \int h_T(t-\tau)p(\tau)d\tau$$

In addition, at any time t, the mean square <u>deviation</u> of the voltage developed across the load from its <u>average</u> value is

$$<[v(t)_{\text{load}} - v_{\text{load average}}(t)]^2> = \frac{\eta}{h\nu} e^2 \int h_T^2(t-\tau)p(\tau)d\tau \tag{5.5}$$

This mean square deviation is a crude measure of the randomness of the load voltage.

In equation (5.5) we have considered the randomness in the load voltage due to the random photodiode hole-electron pair generation times. The amplifier output voltage depends upon this voltage plus noise from the diode biasing resistor (Johnson noise) and from the amplifier internal noise sources. Thus, we can write the amplifier output voltage as follows

$$<v_{\text{amp output}}(t)> = \frac{\eta e}{h\nu} \int h_S(t-\tau)p(\tau)d\tau \tag{5.6}$$

$$<(v_{\text{amp output}} - <v_{\text{amp output}}>)^2> =$$

$$[\frac{\eta e^2}{h\nu} \int h_S^2(t-\tau)p(\tau)d\tau] + N_{\text{Thermal}}^2$$

where $h_S(t)$ is the convolution of $h_T(t)$ with the amplifier impulse response and $N_{thermal}^2$ is the mean squared thermal noise at the amplifier output due to the biasing resistor and the internal amplifier noise sources. It should be noted that this output thermal noise variance depends upon the amplifier type being used and upon the frequency response of any filters included in the amplifier. This noise is often assumed to be proportional to the amplifier bandwidth. However, as we shall see in the next lecture, that is not necessarily true. Let us assume for the moment that the amplifier is fixed so that $N_{Thermal}^2$ is fixed.

From (5.4) and (5.6), we see that for a constant $p(t)=P_o$ the signal-to-noise ratio defined as the square of the amplifier output signal divided by the mean square deviation from the average is given by

$$\frac{<V_{amp\ out}>^2}{<(V_{amp\ out}-<V_{amp\ out}>)^2>}=SNR=\frac{\left(\frac{ne}{h\nu}P_o\int h_S(t)dt\right)^2}{\frac{ne^2}{h\nu}P_o\int h_S^2(t)dt+N_{Thermal}^2} \qquad (5.7)$$

If we now assume $h_S(t)$ is chosen to be the impulse response of an ideal low pass filter with bandwidth B we obtain

$$SNR = \frac{\left(\frac{neP_o}{h\nu}\right)^2}{\frac{2ne^2P_o}{h\nu}B + N_{Thermal}^2} \qquad (5.8)$$

where $N_{Thermal}^2$ depends upon B, the detector load circuit, and the amplifier being used.

We see from (5.8) that for a given thermal noise, $N_{Thermal}^2$ there is a value of P_o = optical input power that gives unity signal-to-noise ratio (average signal = rms "noise" at the amplifier output). This is referred to as the noise-equivalent-power (NEP). The lower the noise equivalent power of a given detector-amplifier combination the less optical power is required to obtain a desired signal-to-noise ratio. In receivers without avalanche gain (to be discussed shortly), and at SNRs near unity, the thermal noise

171

$N^2_{Thermal}$ dominates the "shot noise" (or "quantum noise") in the denominator of (5.8). Thus, the noise equivalent power varies with the receiver bandwidth as the square root of $N_{thermal}$ varies. Since $N_{thermal}$ is <u>sometimes</u> proportional to the receiver bandwidth, NEP is often expressed as "watts" per <u>root Hz</u>). However, unless the amplifier and load resistance are specified, NEP is not a meaningful parameter.

9. <u>Noise in Photodiodes</u> (with avalanche gain)

In order to obtain a bigger detector output signal to combat amplifier and load thermal noises, avalanche gain can be used. Referring to Figure 5.16, the voltaged developed across the load by the displacement currents flowing in the

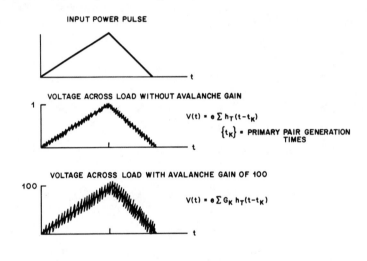

FIG. 5.16: EFFECT OF AVALANCHE GAIN ON DETECTOR STATISTICS

detector is given by

$$v_{load}(t) = e \sum_{n=1}^{N} G_n h_T(t-t_n) \qquad (5.9)$$

In equation (5.9) t_n is the generation time of <u>primary</u> hole electron pair n. G_n is the number of secondary pairs (including the primary) produced through the collision ionization mechanism. For each n, G_n is a random variable with mean G. All the G_n are assumed to be statistically independent. The probability distribution of the random gain G_n depends upon the type of avalanche detector. In particular, it is a function of the ratio of hole collision ionization probability to electron collision ionization probability. That is, as holes and electrons drift through the high field region of the detector, one carrier has a stronger probability per unit length of drift to produce a new hole-electron pair. The derivation of the detector statistics is too complicated to reproduce here. We shall only summarize a few important results.

It can be shown that for "good" multiplication statistics (minimum randomness) primary hole electron pairs should be generated outside of the high field region, so that the more strongly ionizing carrier drifts into the multiplication region.

If the ratio of ionization probabilities is k, then the mean square number of secondaries is related to the mean number of secondaries (mean gain) as follows

$$\langle G_n^2 \rangle = F(\langle G_n \rangle) \cdot \langle G_n \rangle^2 = F(G) \cdot G^2$$

$$(5.10)$$

$$F(G) \cong kG + (2 - \frac{1}{G})(1-k)$$

For an ideal (deterministic) multiplication mechanism we would have $F(G) = 1$. Thus F is a measure of the degradation due to the randomness of the multiplication. Figure 5.17 shows curves of $F(G)$ vs G for various values of k. For good

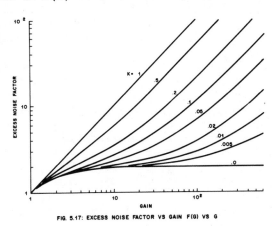

FIG. 5.17: EXCESS NOISE FACTOR VS GAIN F(G) VS G

silicon detectors k is between .02 and .03. From (5.10) and from our previous results in Section 8, it can be shown that the amplifier output voltage has the following statistical properties (see Figure 5.16).

$$\langle v_{amp\ out}(t) \rangle = \frac{eG\eta}{h\nu} \int h_s(t-\tau)p(t)dt$$

$$\langle (v_{amp\ out}(t) - \langle v_{amp\ out}(t) \rangle)^2 \rangle = \frac{e^2 F(G)G^2\eta}{h\nu} \int \qquad (5.11)$$

$$h_s^2(t-\tau)p(\tau)d\tau + N_{thermal}^2$$

Thus, for constant optical input power P_o, and an ideal lowpass $h_s(t)$ with bandwidth B the signal-to-noise ratio is given by the formula (which is analogous to (5.8)).

$$SNR = \frac{\langle V_{amp\ out} \rangle^2}{\langle (V_{amp\ out} - \langle V_{amp\ out} \rangle)^2 \rangle} = \frac{\left(\dfrac{\eta e G P_o}{h\nu} \right)^2}{\dfrac{2\eta e^2 G^2 F(G) P_o B}{h\nu} + N_{thermal}^2}$$

We see that for fixed thermal noise $N_{thermal}^2$ if we set the SNR to unity, there is some value of mean gain G which minimizes the required optical power P_o. That is, the NEP of the avalanche detector-amplifier combination can be minimized by properly adjusting the mean avalanche gain. If the avalanche gain is increased, then the average signal increases relative to the thermal noise. However, if the avalanche gain is increased too far, the first term in the denominator of (5.12) (randomly multiplied shot noise) begins to dominate. Beyond this point, further increases in the average gain make things worse (because F increases). Figure 5.18 shows a typical plot of NEP vs G for a realistic amplifier and detector, assuming a bandwidth requirement of 25MHz. The curve called "Quantum Limit" corresponds to a situation where either $N_{Thermal}^2 = 0$ or $F(G) = 1$ and $G \to \infty$. One should note that the value of G which minimizes the noise equivalent power is not necessarily the optimal value of G to use at high signal-to-noise ratios.

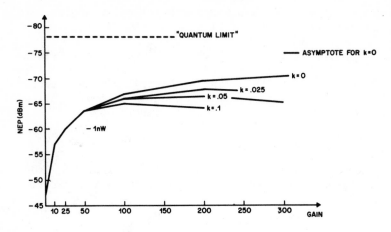

FIG. 5.18: NOISE EQUIVALENT POWER VS. AVERAGE GAIN
TYPICAL 25 MHz RECEIVER

Figure 5.19 shows the required optical power vs G for various
signal-to-noise ratios. Unity SNR is, of course, the NEP

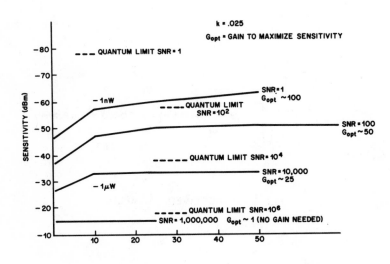

FIG. 5.19: SENSITIVITY IS GAIN TYPICAL 25 MHz RECEIVER

175

curve again. We see that the higher signal-to-noise ratios require less gain. This follows from (5.12).

Equations (5.10-5.12) give information about "second moment" properties of the avalanche gain and the amplifier output signal. For the design of digital communication, more detailed statistical information is needed. This will be discussed further in the next lecture.

10. Example: Designing a Simple Optical Receiver

Suppose we wish to build a receiver for analog modulated light at wavelength .9 μm. The received power waveform is

$$p(t) = P_O [1 + \gamma m(t)] \tag{5.13}$$

where P_O is the average received power, and $m(t)$ is the message. We assume $m(t)$ has peak value unity and that the modulation index, γ , is less than 1. Furthermore, the message $m(t)$ has bandwidth B equal to 1MHz. The receiver is shown in Figure 5.20. It consists of a silicon APD with a 50 ohm amplifier in series. This amplifier provides the

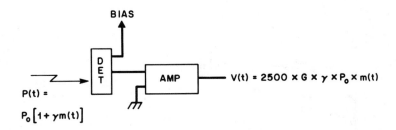

FIG. 5.20: RECEIVER EXAMPLE

176

50 ohm dc resistance to complete the biasing circuit. It should be pointed out that a 50 ohm amplifier is far from optimal in the application from the point of view of mini- mizing thermal noise. However, for simplicity we use it in this example. We shall assume that the detector has a junction capacitance of 5pf and that the amplifier has a shunt capacitance of 5pf. The equivalent circuit is shown in Figure 5.21. For this example, we shall characterize the amplifier noise in terms of its noise figure,-which we

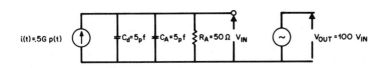

FIG. 5.21: EQUIVALENT CIRCUIT

shall take as 3dB at this bandwidth (1MHz). Thus the ampli- fier has <u>internal</u> noise sources which produce as much output thermal noise as that of the Johnson noise of the load resistance (50 ohms). We shall assume that the amplifier has a gain of 100. We shall assume that the photodiode response is fast enough so that the average diode output current waveform follows the 1MHz input power variations.

We observe that the total load impedance is 50 ohms for frequencies well beyond 1MHz (it rolls off at high frequen- cies). We further observe that the voltage produced across the load neglecting noise is

$$v_{load} = \frac{neG}{h\nu} \cdot 50 \cdot [p(t)] \qquad (5.14)$$

We shall assume that the diode responsivity ($\eta eG/h\nu$) at this wavelength is about (.5 amp/watt) x (avalanche gain).

From our results of Sections 7 and 8, we have the amplifier output average voltage and average noise given by (Average noise is the noise given by Equation (5.11) with p(t) set to its average value P_0).

$$<v_{amp\ output}(t)> = \frac{\eta eG}{h\nu} \cdot 5000 \cdot \gamma \cdot m(t) \cdot P_o$$

$$(5.15)$$

$$\sigma^2 = <(v_{amp\ output}(t) - <v_{amp\ out}(t)>)^2 = \frac{2\eta e^2 GF(G)}{h\nu} (5000)^2 P_o B +$$

$$\frac{8kTB}{50} (5000)^2$$

where
$\quad$ kT = Boltzman's const $\cdot$ Abs Temp ~ 4.15×10^{-21}
$\quad$ B = Bandwidth = 1 MHz = 10^6

In Equation (5.16) we have omitted the DC component of the average output signal.

Now, assume that we wish to have a peak signal to noise ratio of 10^4. That is, we wish to have

$$\frac{\left(\frac{\eta eG}{h\nu} \cdot 5000 \cdot \gamma \cdot P_o\right)^2}{\sigma^2} = 10^4 \qquad (5.16)$$

If we know γ, G and F(G) we can solve for the required average optical power P_0. If we assume that the detector ionization ratio is .025, we can plot $P_{optical}$ required vs G as in Figure 5.22. We see that the optimal avalanche gain for $\gamma = .50$ is approximately 100 and that the required power is -37.5dBm at optimal gain. We also see that the penalty for not using an avalanche detector (G=1) is 17.5dB.

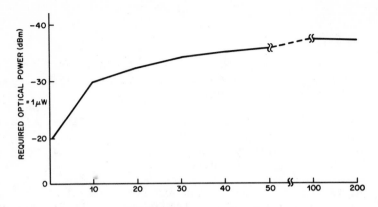

FIG. 5.22: REQUIRED OPTICAL POWER VS GAIN

Note too, that the amplifier output voltage is about
.025 volts. It should be emphasized that in this example,
two important simplifications were in effect. First, the
amplifier was chosen to be a simple 50 ohm input impedance
type, which considerably simplified the calculation of the
thermal noise at the amplifier output. Further, the
performance was specified in terms of the output signal to
noise ratio. This is appropriate for analog modulation
systems. However, for digital systems, error rate is the
criterion of performance. As we shall see in the next
lecture, the correlation between error rate and signal-to-
noise ratio is not always straightforward in optical fiber
systems.

REFERENCES[*]

A. Detectors General

1. H. Melchior, M. B. Fisher and F. Abrams, "Photodetectors for Optical Communication Systems", Proc. IEEE, Vol. 58, pp. 1466-1486, October, 1970.

2. L. K. Anderson, M. DiDomenico and M. B. Fisher, "High Speed Photodetectors for Microwave Demodulation of Light", In Advances in Microwaves, Vol. 5, L. Young Ed., New York, Academic Press, 1970.

3. H. Melchior, "Demodulation and Photodetection Techniques", in Laser Handbook, F. T. Arecchi and E. D. Schulz-Dubois, Eds., Amsterdam Netherlands, Elsevier, North Holland, pp. 628-739.

B. Avalanche Detector Statistics

4. R. J. McIntyre, "Multiplication Noise in Uniform Avalanche Diodes", IEEE Trans. on Electron Devices, Vol. ED-31, pp. 164-168, January, 1966.

5. R. J. McIntyre and J. Conradi, "The Distribution of Gains in Uniformly Multiplying Avalanche Photodiodes", IEEE Trans. Electron Devices, Vol. ED-19, pp. 713-718, June, 1972.

6. S. D. Personick, "New Results on Avalanche Multiplication Statistics with Applications to Optical Detection", BSTJ, Vol. 50, pp. 167-189, January, 1971.

7. S. D. Personick, "Statistics of a General Class of Avalanche Detectors with Applications to Optical Communication", BSTJ, Vol. 50, pp. 3075-3095, December, 1971.

8. P. P. Webb, R. J. McIntyre and J. Conradi, "Properties of Avalanche Photodiodes", RCA Review, pp. 234-278, June, 1974.

---

[*]Due to the tutorial nature of this material, references are given by topic rather than to specific results.

References - 2

C. Applications of Photodiodes in Systems

9. Chapter 6 of this text.

10. W. M. Hubbard, "Efficient Utilization of Optical Frequency Carriers for Low and Moderate Bandwidths", BSTJ, Vol. 52, pp. 731-765, May-June, 1973.

Chapter 6 - Design of Repeaters for Fiber Systems

S. D. Personick

Bell Telephone Laboratories
Holmdel, NJ   07733

1. Introduction

In the previous chapter we described the process of
photodetection,-conversion of optical power into an electrical current. We reviewed the physical mechanism, and
the mathematical model for that mechanism both for PIN
diodes and avalanche photodiodes. We are now ready to
describe the theory and practice of repeater design.

A repeater consists of a detector, an amplifier,
equalizer, and a regenerator, - followed by a driver-
optical source pair as shown in Figure 6.1. Once the
signal has been amplified, the rest of the processing,

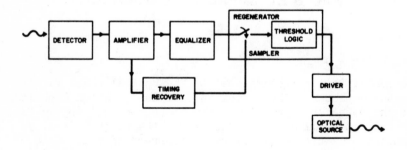

FIG 6.1: OPTICAL REPEATER

ne driver, is fairly conventional,-that is it is
cal to the processing done in conventional wire system
ters. Although we shall review the total repeater
gn, we shall be particularly interested in the low noise
amplifier which must follow the photodetector. Thus, we
begin with a review of noise in baseband amplifiers.

2. Amplifiers for Capacitance Sources

We entitle this section "amplifiers for capacitive
sources" because the photodiode equivalent circuit is
essentially a current source in parallel with a capacitor.
This is shown in Figure 6.2, and follows from the description
in Chapter 5. Neglecting shot noise for the moment, the

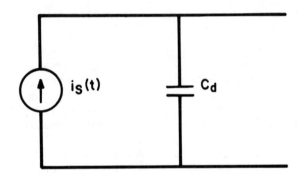

FIG. 6.2: CAPACITIVE SOURCE

current $i_s(t)$ is proportional to the optical power falling
on the detector. The capacitance is the junction capacitance
of the detector, typically a few pf (but conceivably a fraction
of a pf in small area photodiodes). The proportionality
constant relating milliamperes of current to milliwatts of
power is the detector responsivity, - typically .5 multiplied
by the avalanche gain if any, (for silicon detectors at
.9 μm wavelength).

184

Schematically, the amplifier can be modeled as shown in Figure 6.3. Note the following features. The amplifier has an input impedance consisting of a capacitance C_a in

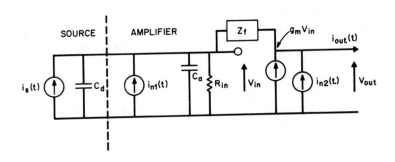

FIG. 6.3: AMPLIFIER SCHEMATIC

parallel with a resistance R_{in}. This resistance may be large or small depending upon the type of device being used (more to follow on this point). The capacitance will generally be a few pf, - although smaller values are possible with special device design. {The amplifier gain is represented by the voltage controlled current source g_m. For field effect transistors g_m is fixed. Typically it is about 1000-5000 micro-Siemens for silicon FETs. For bipolar transistors g_m is the current gain β divided by the transistor input resistance r_{in}, (neglecting base series resistance). FET and Bipolar examples are shown in Figure 6.4. These are just examples, one is not constrained to use the common emitter or common source configuration. Getting back to Figure 6.3, there is included a feedback impedance Z_f. This can be stray capacitance or it can be a lumped component intentionally

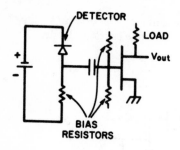

FIG. 6.4 a : FET AMPLIFIER

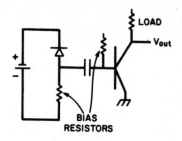

FIG. 6.4 b: BIPOLAR AMP

included by the designer. In order to limit the scope of this discussion, we shall assume that the feedback impedance is infinite. The reader who wishes to do a more comprehensive study of amplifier designs is referred to the literature. We shall also restrict our discussion here to a single stage of amplification. Extension of the ideas to multistage amplifiers is also left to the literature.

Once again referring to Figure 6.3, we observe two noise sources. The noise sources are assumed to be independent white Gaussian noise sources with spectral heights I_{n1} and I_{n2} amps^2 per Hz. For both the FET and the bipolar transistor, $i_{n1}(t)$ is the shot noise associated with the gate or base (respectively) current plus Johnson noise from any biasing resistors at the device input. In the FET, $i_{n2}(t)$ is associated with the resistance of the source-drain channel. In the bipolar transistor $i_{n2}(t)$ is shot noise associated the collector bias current.

In designing an amplifier, our aim is to choose a device and pick an operating point (bias condition) so as to amplify the detector output signal with as little added noise as possible. We begin by calculating both the signal and the added amplifier noise at the amplifier output.

First, we observe that the signal at the amplifier output is given as follows (in the frequency domain)

$$I_{out}(\omega) = F\{i_{out}(t)\} = I_s(\omega) \left[\frac{1}{j\omega(C_a+C_d) + \frac{1}{R_{in}}} \right] g_m \quad (6.1)$$

Note that the gain vs frequency of the amplifier is not constant. This is because of the frequency "rolloff" of the total impedance shunting the signal $i_s(t)$. We could place a small resistance in shunt with the amplifier input, but that would add extra Johnson noise. Instead, we shall include a frequency "rollup" circuit at the amplifier output as shown in Figure 6.5. The signal at the output of this "equalizer" is given by

$$v_{eq\ out}(\omega) = I_s(\omega)Z \quad \text{for } \omega \leq 2\pi B \quad (6.2)$$

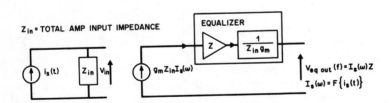

FIG. 6.5: AMPLIFIER-EQUALIZER

187

Note that the equalization takes place after further amplification of the signal. The noise spectral density at the equalizer output is given by (referring to Figure 6.5)

$$
N_{eq\ out}(\omega) = I_{n1} Z^2 + \frac{I_{n2} Z^2}{g_m^2} \left[\omega^2 (C_a + C_d)^2 + \left(\frac{1}{R_{in}} \right)^2 \right] \quad (6.3)
$$
$$
\text{for } \omega < 2\pi B
$$

We see that at each frequency, the noise is minimized if g_m is large, the total capacitance $(C_a + C_d)$ at the amplifier input is small, and the spectral heights I_{n1} and I_{n2} are small. Unfortunately, these parameters are not all independent and at the control of the amplifier designer. In order to see the tradeoffs, we shall study the FET and the Bipolar amplifiers separately.

2.1 FET Amplifiers

For FET amplifiers, I_{n1} is typically negligible, at least at bit rates about 0.1Mb/s. To simplify our discussion we shall set it to zero. Also, we shall set R_{in} to infinity which is also appropriate above 0.1Mb/s. I_2 in an FET is given by 2.8 kT g_m. Where kT is Boltzmanns constant · absolute temperature (about 4.3 x 10^{-21}. From (6.3) we obtain

$$
N_{eq\ out}(\omega) = 2.8 kT\ Z^2 \Big/ \left[g_m / (C_a + C_d)^2 \right] \quad (6.4)
$$

We see, therefore, that for an FET, $g_m/(C_d + C_a)^2$ is a figure of merit with regard to noise performance. A given FET has a fixed g_m and C_a. However, conceivably, FETs could be designed to optimize this figure of merit. For a given material (say silicon) g_m and C_a tradeoff in the following fashion

$$
\frac{g_m}{C_d} \sim \text{constant} \quad (6.5)
$$

Thus, to optimize the figure of merit $g_m/(C_d + C_a)^2$, one should have $C_a = C_d$. In current fiber system art, FETs are not specifically designed to match the detectors.

2.2 Bipolar Amplifiers

Analysis of the bipolar amplifier is somewhat more complicated than the FET analysis. First, we must observe that the value of R_{in} (see Figure 6.5) the amplifier input resistance is under the control of the designer. R_{in} is given by

$$R_{in} = \frac{kT}{eI_{in\ bias}} \qquad (6.6)$$

where $I_{in\ bias}$ is the base bias current.

Next observe that I_{n1}, the spectral height of the noise source $i_{n1}(t)$, is given by $2eI_{in\ bias} = 2kT/R_{in}$. Also, the g_m of the transistor is β/R_{in}, where β is fixed and typically about 100. Finally, the spectral height I_{n2} of noise source i_{n2} is $2e\beta I_{in\ bias} = 2kT\beta/R_{in}$. Substituting into Equation (6.3) we get

$$N_{eq\ out}(\omega) = \frac{2kT}{R_{in}} Z^2 + \frac{2kT\ R_{in} Z^2}{\beta} \left[\omega^2 (C_a + C_d)^2 + \frac{1}{R_{in}^2} \right] \quad (6.7)$$

Since R_{in} is a parameter which we can control, we can optimize (6.7) to obtain the lowest noise at any particular frequency ω. Also, it can optimize R_{in} to minimize the total noise in some frequency band B. For a given bandwidth B, the optimal R_{in} and minimized noise are given by

$$R_{in\ optimal} = \frac{1}{2\pi(C_a + C_d)B} \sqrt{3\beta} \qquad \text{for } \beta \gg 1$$

$$N_{eq\ out\ total}^{(B)} = \frac{1}{2\pi} \int_0^{2\pi B} N_{eq\ out}(\omega)d\omega \Bigg|_{opt\ R_{in}} = \qquad (6.8)$$

$$\left[\frac{4kT(2\pi[C_a + C_d])}{\sqrt{3\beta}} \right] B^2 Z^2$$

From (6.8) we see that $\beta/(C_d+C_a)^2$ is a figure of merit for bipolar transistors. Note that at optimal bias, the total input impedance shunting the detector rolls off with increasing frequency. Thus, the equalizer does not become superfluous. Also note that if the amplifier is kept optimized, then the total noise in bandwidth B is proportional to B^2. (For the FET, it is proportional to B^3 as can be seen from equation (6.4)).

2.3 Summary and Further Comments

We have studied amplifier and amplifier noise with the aim of designing an amplifier which adds as little noise as possible to the detector output signal. In practical systems, one sometimes trades off increased amplifier noise for other desirable features such as large dynamic range or convenience of using existing commercial amplifiers. We assumed that a frequency rollup equalizer was used to compensate for the frequency rolloff at the amplifier input (due to the capacitive impedance). Other approaches, in particular feedback, can achieve similar results. In the literature such feedback amplifiers are referred to as "transimpedance" amplifiers. The advantage of these transimpedance amplifiers is increased dynamic range. However, at high frequencies, they are difficult to implement. The interested reader is referred to the growing literature on this subject describing both theory and practice in amplifier design.

If the reader carries away one thought from the previous discussion, it should be that standard 50-ohm-input-impedance amplifiers add far more noise than will be added by an amplifier specifically designed to work with a photodetector.

3. The Linear Channel

In the chapter on detectors, we showed that over a reasonable (but finite) range of optical illumination, the current that flows in the detector is on the average proportional to the optical power impinging upon it. Figure 6.6 shows what may be called the linear channel of a repeater. As in the detector, the output of each block is a linearly filtered version on the input to that block. Neglecting noise for the moment, we see that the linear channel serves two purposes: amplification and equalization.

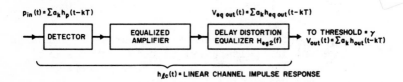

FIG. 6.6: LINEAR CHANNEL

3.1 Equalization

The light falling upon the detector is assumed to be a sum of pulses

$$p_{in}(t) = \sum a_k h_p(t-kT) \qquad (6.9)$$

where $a_k = 0$ of 1 for each k.

In (6.9) the individual pulses may overlap, due to pulse spreading as the light propagates along the fiber. After detection and amplification, we obtain

$$v_{eq\ out}(t) = \sum a_k h_{eq\ out}(t-kT) \qquad (6.10)$$

where $h_{eq\ out}(t)$ includes the amplifier equalizer, - which compensates for any integration at the amplifier input. The pulses $h_{eq\ out}(t)$ may overlap just as the input power pulses may. This is referred to as intersymbol interference. If $v_{eq\ out}(t)$ were sampled once per time slot, then the resulting sample values would depend upon more than one digit (and noise). If one knows the shape $h_{eq\ out}(t)$, (which requires knowledge of $h_p(t)$) then one can build an equalizer filter to reduce the pulse overlap. If the desired output pulses are $h_{out}(t)$ and the equalizer input pulses are $h_{eq\ out}(t)$, then the equalizer frequency response must be

$$H_{eq\ 2}(f) = \frac{F\{h_{out}(t)\}}{F\{h_{eq\ out}(t)\}} \tag{6.11}$$

Since $|H_{equalizer\ 2}(f)|$ is in general an increasing function of f (for large initial pulse overlap) noise enhancement occurs. That is, noise entering this second equalizer is increased in mean square value. Thus there is some tradeoff between reduced intersymbol interference and enhanced noise. As a practical matter, it is tempting to leave out the equalizer ($H_{eq\ 2}$) and accept the intersymbol interference, - since inclusion of this equalizer may require more information about the fiber impulse response than is available to the repeater designer.

Example: Suppose the received optical pulses are Gaussian in shape as shown in Figure 6.7.

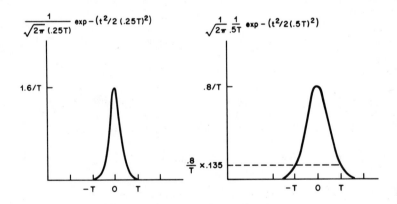

FIG. 6.7: GAUSSIAN PULSES

192

$$h_p(t) = \frac{1}{\sqrt{2\pi(.25T)^2}} \; e^{-t^2/2(.25T)^2} \qquad (6.12)$$

where T is the width of a time slot.

Suppose next that after passing through the amplifier we obtain the following individual pulse shape (see Figure 6.7)

$$h_{eq\ out}(t) = \frac{1}{\sqrt{2\pi(.5T)^2}} \; e^{-t^2/2(.5T)^2} \qquad (6.13)$$

The pulses given in (6.13) overlap. As a measure of intersymbol interference, we can plot what is referred to as an eye diagram. It consists of two curves. The first is a plot of the <u>largest</u> value $v_{out}(t)$ can take when $a_O = 0$ and the other a_k can be chosen arbitrarily (see equation (6.10) and Figure 6.6). The second curve is a plot of the <u>smallest</u> value $v_{out}(t)$ can assume if $a_O = 1$ and the other a_k are again chosen arbitrarily. The eye is plotted in Figure 6.8 for the pulse shape of (6.13). Referring to Figure 6.8, the value

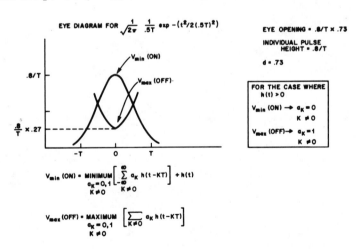

FIG. 6.8: EYE DIAGRAM

d = (peak eye opening / individual pulse height) is called the fractional eye opening. It is a measure of how much the ability to distinguish between a_o = 0 and a_o = 1 has been degraded.

Once again, we could in principle build an equalizer to modify the individual pulse shape resulting in a larger fractional eye opening. However, this enchances the noise, producing a tradeoff.

4. Calculating the Error Rate

Referring to Figure 6.6, we see that the decision process consists of sampling the linear channel output once per time slot. If the sample value exceeds some threshold, we decide that a_k = 1 (see Equation 6.8) if the sample is below the threshold, we decide that a_k = 0. The probability of error is the chance that $v_{out}(kT)$ exceeds threshold and a_k = 0 or $v_{out}(kT)$ is below threshold and a_k = 1. In order to calculate this error probability, we would have to know the probability distribution of $v_{out}(t)$ under both hypotheses (a_k = 0, a_k = 1). In receivers with avalanche gain, it is difficult to make this calculation exactly. Various approximation techniques are available. The simplest is the Gaussian approximation, - which we shall outline next. The interested reader is referred to the evolving literature for more detailed information and alternative approximations.

4.1 The Gaussian Approximation

Consider the output of the linear channel shown in Figure 6.6. At sampling time kT, this output is given by

$$v_{out}(t)\Big|_{t=kT} = \sum_j a_j h_{out}(t-jT) + n(t)\Big|_{t=kT} \qquad (6.14)$$

where the noise n(t) consists of two parts: amplifier thermal noise and randomly multiplied shot noise associated with the detection process. The thermal noise was discussed in Section 2 above. The randomly multiplied shot noise was discussed in Chapter 5. Recall that the detector output consists of electrons which are created at random times, but at an average rate proportional to the incident optical power. The average current emitted by the detector is thus proportional to the incident optical power. The deviations from this average current (due to the randomness of when and how many electrons will be emitted) results in the randomly multiplied shot noise. In the Gaussian approximation, we pretend that $n(kT)$ is a Gaussian random variable. With this approximation, we need only know the mean and variance of $n(kT)$. The thermal noise has zero mean value, and the randomly multiplied shot noise has mean value also equal to zero (the mean value of the current is included in the signal term in (6.14)). Thus $n(kT)$ is completely characterized by its variance under this approximation. We obtain the following two expressions from Figure 6.6, and the statistical model of a Poisson process discussed in Chapter 5.

$$< v_{out}(kT)> = v_{out\ average}(kT) = \frac{e\eta \overline{G}}{h\nu} \int P_{in}(t')h_{lc}(t-t')dt' \Big|_{t=kT}$$

$$= \sum a_j h_{out}(t-kT) \Big|_{t=kT}$$

$$<n^2(kT)> = N^2_{av}(kT)$$

$$= N^2_{amplifier} + \int P_{in}(t')\frac{e^2 \eta}{h\nu} \overline{G^2} h^2_{lc}(t-t')dt' \Big|_{t=kT}$$

$$= \frac{e^2 \eta G^2}{h\nu} \sum a_j \int h_p(t')h^2_{lc}(t-t')dt' \Big|_{t=kT} + N^2_{amplifier}$$

where $\frac{e\eta}{h\nu}$ = detector responsivity not including avalanche gain

e = electron charge, $\overline{G}$ = average avalanche gain,

$\overline{G^2}$ = mean square avalanche gain

$h_{lc}(t)$ = linear channel impulse response

$N^2_{amplifier}$ = amplifier contribution to output noise

From (6.15) we see that in general, both the signal $v_{out\ average}(kT)$ and the noise $N^2_{av}(kT)$ depend upon the complete set of symbol values (a_j). As a practical matter, this dependence is very weak for $|j-k|$ larger than some

195

small integer. Thus, it is possible to calculate the signal and noise variance for all possible symbol combinations which affect these two quantities. For each combination of $\{a_j\}$ for $j \neq k$, the probability of error is given by

$$P_f = \text{Prob}(v_{out}(kT) > \text{THRESHOLD}) \text{ if } a_k = 0, \{a_j\} \text{ given for } j \neq k$$

$$\text{or} \qquad (6.16)$$

$$P_m = \text{Prob}(v_{out}(kT) < \text{THRESHOLD}) \text{ if } a_k = 1, \{a_j\} \text{ given for } j \neq k$$

Since we have assumed that $n(kT)$ is approximately a Gaussian random variable, these probabilities reduce to

$$P_f = \text{ERFC*}\left[\frac{\gamma - [v_{out\ av}(kT)| \{a_j\}, a_k = 0]}{\sqrt{(N_{av}^2(kT)|\{a_j\}, a_k = 0}} \right]$$

$$(6.17)$$

$$P_m = \text{ERFC*}\left[\frac{[v_{out\ av}(kT)| \{a_j\}, a_k = 1] - \gamma}{\sqrt{N_{av}^2(kT)|\{a_j\}, a_k = 1}} \right]$$

where

$$[x| \{a_j\}, a_k] \triangleq \text{ particular value } x \text{ assumes for the given set of } \{a_j\} \text{ and } a_k$$

and

$$\text{ERFC*}(x) = \frac{1}{\sqrt{2\pi}} \int_0^\infty e^{-y^2/2} dy \qquad (6.18)$$

ERFC* is a tabulated function, a graph of which is shown in Figure 6.9.

Using (6.17) one can find the optimal threshold, avalanche gain and required optical power to achieve a desired error rate. One can calculate the effects of pulse spreading in the fiber on the receiver sensitivity. (Required, optical power to achieve a desired error rate.)

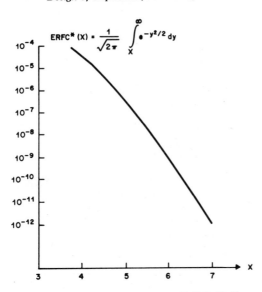

FIG. 6.9: ERROR FUNCTION COMPLEMENT VS X

The reader is cautioned however, that the Gaussian approximation is not completely satisfactory in systems employing avalanche detectors. The interested reader should consult the literature for more tedious, but more accurate analysis techniques. Laboratory experience shows that the Gaussian approximation is adequate for the calculation of receiver sensitivity (to within a dB or so). One shortcoming which has appeared in practice is that in systems with avalanche gain, the optimal threshold turns out to be closer to the center of the "eye" than the Gaussian approximation would predict. This is a consequence of the highly skewed nature of the probability distribution of randomly multiplied shot noise.

5. Computational Results on Error Rate

Based on the Gaussian approximation, receiver sensitivity curves have been calculated for receivers employing bipolar or FET front ends and for receivers with and without avalanche gain. Assuming no significant pulse spreading in the fiber, the results are shown in Figure 6.10. In that figure, a 10^{-9} error rate is assumed. The rather large curve widths take into account component ranges. Crosses on the curves are for experimental systems which have been reported. For example, at 50 Mb/s, a receiver sensitivity of -57dBm of optical power

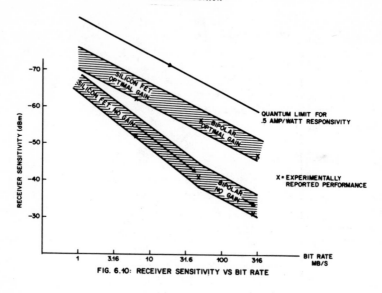

FIG. 6.10: RECEIVER SENSITIVITY VS BIT RATE

was achieved.   Thus, for a transmitted power level of 1mW,
57 dB of fiber loss could be allowed, - provided pulse
spreading is sufficiently small.

The effects of pulse spreading have also been calculated
for typical systems.   It has been found that pulse spreading
is negligible if the impulse response of the fiber $h_{fiber}(t)$
satisfies

$$\sigma < .25T = .25 \text{ pulse spacing} \tag{6.19}$$

where

$$\sigma^2 = \frac{\int t^2 h_{fiber}(t)\, dt}{\int h_{fiber}(t)\, dt} - \left[\frac{\int t h_{fiber}(t)\, dt}{\int h_{fiber}(t)\, dt} \right]^2 \tag{6.20}$$

The parameter σ is referred to as the rms pulse
width.   For σ greater than .5T, intersymbol interference
becomes severe, and the receiver sensitivity degrades
rapidly (either the eye closes, - or for equalizing
receivers, noise enhancement increases rapidly).

198

If the fiber impulse response satisfies (6.19) then the allowable fiber loss is the difference between the transmitted power and the receiver sensitivity. Under those circumstances, the receiver is said to be loss limited. If the fiber pulse spreading is excessive, then the fiber length must be smaller than the loss limited value to reduce the intersymbol interference. The receiver is then said to be dispersion limited.

6. Timing Recovery

Timing recovery for sampling the output of the linear channel is accomplished in the same way as for wire system repeaters. Either a crystal filter or a phase locked loop can be used. One interesting difference between the fiber system and the wire system is that the signal in the linear channel may contain a timing component even without rectification. This is because the optical system uses "on-off" signaling rather than plus-minus signaling as is used in wire systems. In order for this timing component to be present, the transmitted signal must be in a "return-to-zero" format. If desired, the linear channel signal can be differentiated and rectified as in a wire repeater in order to obtain an enhanced timing component or if a non-zero-to-return format is used at the transmitter.

7. Driver Circuitry

Optical sources used in fiber systems are high current low impedance devices. Typical operating currents exceed 100ma. Typical impedances are a few ohms. Thus, for efficient operation high current low impedance transistor drivers are needed. At modest modulation speeds (less than 25Mb/s) the optical source will follow the modulation (in a non-linear fashion). At high modulation speeds, the optical components exhibit pattern dependent effects, - where the height of one emitted pulse depends upon whether or not pulses were emitted in previous time slots. Thus, at high modulation rates, fairly sophisticated driver designs are required. Successful modulation of LEDs at up to 50Mb/s and lasers beyond 300Mb/s have been reported.

When laser sources are used, compensation must be included in the driver for the temperature dependent threshold of these devices. The threshold may also change as the device degrades with age. One approach which has been used is to monitor the light emitted by the laser locally with a PIN detector. The detector output is then used in a feedback arrangement to stabilize the output of the laser.

This feedback approach has also been used to linearize light emitting diodes for analog modulation purposes.

8. Alternative Modulation Formats

The previous discussion has been restricted to digital modulation, - since that appears most promising for optical transmission in the near future. Other modulation formats are certainly possible, and have been reported in the literature.

If linearity is not too severe a problem, direct analog modulation of LEDs (or perhaps even lasers) is possible. In such a system, the LED drive current is linearly modulated about some operating point. As mentioned optical feedback can be employed to improve the linearity.

Pulse position modulation may be attractive for use with laser sources which emit narrow pulses of high peak power (but which must operate at low duty cycles). Here linearity depends upon the ability to linearly delay the emission time of a pulse.

Subcarrier frequency modulation has also been proposed for fiber systems. Here, however, the delay distortion in the fiber may be a severe limitation. In subcarrier systems, a sinoisoidal drive current is superimposed upon a bias to form the LED input. The sinusoidal component is then modulated using any FM format.

9. Conclusions

Sensitive receivers for optical fiber systems can and have been designed. Allowable losses between transmitter and receiver of 50dB or more are possible at data rates beyond 50Mb/s. For best performance, the receiver front end amplifier must be designed to work with a capacitive source such as a PIN or avalanche photodiode. The driver must be designed to work efficiently with a high current, low impedance optical source. Beyond this, however, much of the optical repeater circuitry is identical to that of a wire repeater.

S. D. Personick

GENERAL REFERENCES*

A. <u>Amplifier Noise and Design</u> (Theory)

1. A. Van der Ziel, "Noise: Sources, Characterization", Englewood Cliffs, NJ, Prentice Hall, 1970.

2. A. B. Gillespie, "Signal, Noise and Resolution in Nuclear Particle Counters", New York, Pergamon Press, Inc., 1953.

3. S. D. Personick, "Receiver Design for Digital Fiber Optic Communication Systems", BSTJ, July-August, 1973, Vol. 50, No. 1, pp. 843-886.

4. J. E. Goell, "Input Amplifiers for Optical PCM Receivers", BSTJ, November, 1974, Vol. 53, No. 9, pp. 1771-1794.

5.

B. <u>Repeater Design</u> (Experimental)

6. J. E. Goell, "An Optical Repeater with a High Impedance Input Amplifier", BSTJ, Vol. 53, No. 4, April, 1974, pp. 629-643.

7. P. K. Runge, (to be published) "An Experimental 50Mb/s Fiber Optic Repeater", IEEE Trans. on Communications, also "A 50Mb/s Repeater for a Fiber Optic PCM Experiment", Proceedings of ICC'74 (Minneapolis).

8. J. E. Goell, "A 275Mb/s Optical Repeater Experiment Employing a GaAs Laser", Proc. IEEE (lett.), Vol. 61, pp. 1504-1505, October, 1973.

9. T. Ozeki and T. Ito, "A 200Mb/s PCM DH GaAlAs Laser Communication Experiment", Presented at 1973 Conference on Laser Engineering and Applications, Washington, DC, May, 1973; abstract in IEEE J. Quantum Electronics, Vol. QE-9, p. 692, June, 1973.

---

*Due to the tutorial nature of this material, references are given by topic rather than to specific results.

General References - 2

10. F. M. Banks, et al, "An Experimental 45Mb/s Digital Transmission System Using Optical Fibers", Proc. of ICC'74, Minneapolis, MN.

11. T. Uchida, et al, "An Experimental 123Mb/s Fiber Optic Communication System", Proc. Topical Meeting in Optical Fiber Transmission, Williamsburg, VA, January 7-9, 1975.

C. Signals and Noise (Theory)

12. R. W. Lucky, J. Salz and E. Weldon, "Principles of Data Communication", New York, McGraw Hill, 1968.

13. W. Davenport and W. Root, "Random Signals and Noise", New York, McGraw Hill, 1958.

D. Optical Drivers

14. M. Chown, et al, "Direct Modulation of Double Hetero-structure Lasers at Rates up to 1Gb/s", Elec. Letts., Vol. 9, pp. 34-36, January, 1973.

15. P. K. Runge, (to be published) "An Experimental 50Mb/s Fiber Optic Repeater", IEEE Trans. on Communications, also "A 50Mb/s Repeater for a Fiber Optic PCM Experiment", Proceedings of ICC'74 (Minneapolis).

E. Alternative Modulation Schemes

15. W. S. Holden, "Pulse Position Modulation Experiment for Optical Fiber Transmission", Proc. Topical Meeting on Optical Fiber Transmission, January 7-9, 1975, Williamsburg, VA.

16. A. Szanto and J. Taylor, "An Optical Fiber System for Wideband Transmission", Proceedings of ICC'74, Minneapolis, MN.

F. Repeater Performance Analysis

17. S. D. Personick, "Receiver Design for Digital Fiber Optic Communication Systems", BSTJ, July-August, 1973, Vol. 50, No. 1, pp. 843-886.

General References - 3

18. S. D. Personick, "Statistics of a General Class of Avalanche Detectors with Applications to Optical Communication", BSTJ, Vol. 50, No. 10, pp. 3075-3095, December, 1971.

19. S. D. Personick, "New Results on Avalanche Multiplication Statistics with Applications to Optical Detection", BSTJ, Vol. 50, No. 1, pp. 167-189, January, 1971.

CHAPTER 7 - DESIGN CONSIDERATIONS FOR
MULTITERMINAL NETWORKS

M. K. Barnoski

Hughes Research Laboratories
Malibu, California 90265

1. INTRODUCTION

It is the principal intent of this chapter to utilize
the information presented in the preceding chapters of this
text to address the problems in applying fiber optic transmission lines to distribute information to numbers of
remote terminals, all interconnected in a bidirectional
data distribution system. Such data distribution networks
are of particular interest in the short-range class of
applications which include information transfer within
vehicles such as aircraft and ships and also within struc-
trues such as office buildings, manufacturing facilities,
power plants, and power switching yards. For these types of
application, maximum interterminal spacings will be a few
hundred meters or less, with information rates ranging
from kilohertz to possibly a few hundred megahertz.

In addition to these relatively short-range applications, it is also an interesting exercise to consider the
more long-range possibilities such as the utilization of
fiber optic technology for distribution of data in a two-
way CATV network. Both of these applications utilize the
concept of the data bus, a single transmission line which
simultaneously carries many different multiplexed signals
and serves a number of spatially distributed terminals.
The engineering design considerations which must be
investigated in order to properly design multiterminal
fiber optic networks will now be discussed.

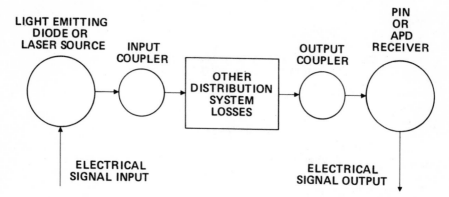

LIGHT EMITTING
DIODE OR
LASER SOURCE   INPUT
                COUPLER

OTHER
DISTRIBUTION
SYSTEM
LOSSES

OUTPUT
COUPLER

PIN
OR
APD
RECEIVER

ELECTRICAL
SIGNAL INPUT

ELECTRICAL
SIGNAL OUTPUT

Fig. 7.1   Components comprising a terminal-to-terminal
section of a multiterminal distribution system.

The components which comprise a typical terminal-to-terminal section of a multiterminal system are illustrated schematically in Fig. 7.1. These are the transmitting LED or laser source, the input and output couplers, the fiber cabling, the other distribution system components (which from a terminal-to-terminal viewpoint represent added loss), and finally, the receiving photodetector. Clearly, the amount of optical power available for distribution in the system depends upon the amount of power launched into the transmission line at any given terminal, the amount lost in the fiber waveguide cabling due to attenuation, and the amount of optical power required at the receiving photodiode necessary to maintain the error rate (signal-to-noise ratio) desired. Thus, the design of a data distribution system dictates that the following set of questions be addressed: (1) What error rate is required?   (2) How much optical power is required incident on the photodetector to maintain the desired error rate?   That is, how good is the receiver?   (3) How much optical power is emitted from the source?   (4) How much optical power is coupled into the transmission line at the input?   (5) How much optical power is available for data distribution?   That is, how high can the distribution system losses be?   (6) What is the most efficient distribution system for your particular application?

2. MINIMUM DETECTABLE RECEIVER POWER

Assuming that the acceptable error rate is one error for every 10^9 bits transmitted, the minimum detectable power required incident on the active surface of the photodetector can be determined as discussed in Chapter 6. For convenient reference, plots of minimum detectable power versus information bit rate, as presented by M. Di Domenico, Jr.,[1] are reproduced here in Fig. 7.2. As can be seen from the figure, the minimum power incident on the receiver necessary to maintain an error rate of 10^{-9} increases with increasing information bit rate. This is true for both silicon PIN and APD (avalanche photodiode) detectors with either FET or bipolar front end preamplifiers. Plots were made for both bipolar and FET input amplifiers. The data points shown are experimental results obtained at bit rates of 6, 50, and 300 Mbit/sec. From the figure it can be seen that, for example, at 50 Mbit/sec data rate, the power required at a receiver employing a PIN photodiode is approximately -50 dBm, while that required if an APD is used is only -60 dBm. These results are valid for the case where the optical pulse width is less than a time slot, that is, in systems where inter-symbol interference is negligible. This is a valid assumption for relatively short length, medium data rate distribution systems currently under consideration for most military applications.[2] This may, of course, not be a valid assumption for long distance, high data rate distribution system, such as two-way CATV networks.

3. COMPATIBLE LIGHT-EMITTING SOURCES

Given the optical power required incident on the receiver, it is necessary next to determine the amount of power coupled into the transmission line at the input. This, of course, depends on the total power emitted from the source. As mentioned in Chapter 3, the two principal opto-electronic sources of interest for use in fiber optic communications systems are the GaAs LED and the GaAs injection laser diode. The total optical output power emitted from both low and high brightness LED's and from cw room temperature injection lasers are plotted as a function of drive current in Fig. 7.3. The curve for the RCA laser was extracted from Fig. 4.12 of Chapter 4, while that for the

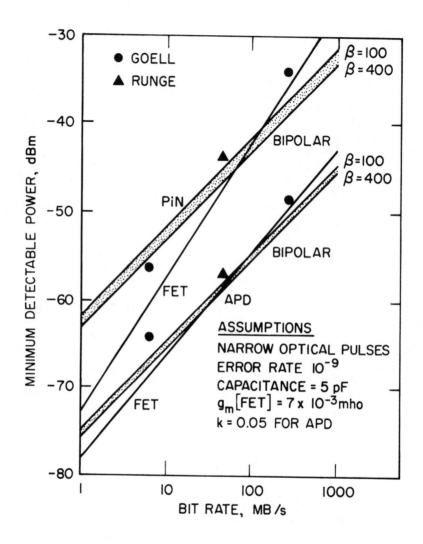

Fig. 7.2.   Minimum detectable optical power plotted as a
function of bit rate.   (After M. Di Domenico,
Jr., Ref. 1.)

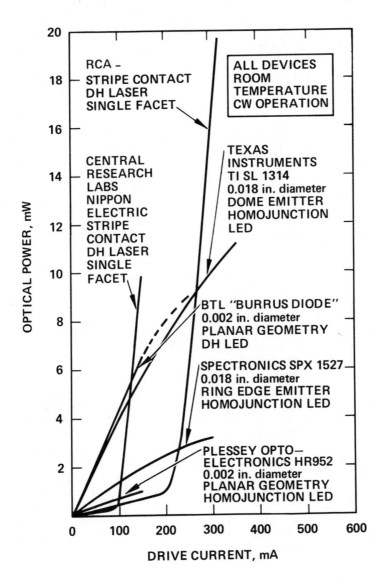

Fig. 7.3. Total optical output power from several different types of LED's and injection lasers as a function of drive current.

Nippon Electric laser diode was obtained from Ref. 3. It should be noted that laser diodes capable of room temperature, cw operation are now becoming commercially available from both RCA and Nippon Electric Company. The optical output power available from a surface-emitting, high brightness, light-emitting diode developed by Burrus[4] is also shown in Fig. 7.3, as is the power emitted by a small emitting area homojunction LED manufactured by Plessey Opto-Electronics.

Also plotted in Fig. 7.3 are output powers obtainable from two commercially available, large area, low brightness LED's. Reference to the figure reveals that laser diodes and high brightness LED's are capable of emitting ∿4 dBm of optical power when driven with 150 to 250 mA of current, while the large area, low brightness LED can emit as much as 10 dBm of optical power at 300 mA drive current.

4. INPUT COUPLING EFFICIENCY

Having established an estimate for the total amount of optical power available from the source, it remains to determine the fraction of this power coupled into the transmission line at the input. The problem of determining the coupling efficiency for both single-strand and multistrand bundles has been treated in detail in Chapter 3. The results of that treatment revealed that since the input coupling loss for coupling power from a small area device into a single strand of fiber waveguide varies as $(NA)^2$, a considerable amount of power emitted by the source is lost in the coupler. For a fiber bundle system, however, it is possible to minimize the input coupling loss by properly choosing the source and the bundle configuration and by using an intervening optical element such as a lens. If the proper choice is made, the input coupling loss can be reduced to that which results from the packing fraction loss. As a result, the packing fraction losses for various bundle configurations will now be discussed.

5. PACKING FRACTION LOSSES

Estimates of bundle packing fraction losses can be obtained by considering some basic bundle configurations,

such as those illustrated in Fig. 7.4. The packing
fraction for a hexagonal, close-packed bundle containing
a total of N fibers with n_d fibers along the diameter of
the smallest circumscribed circle containing all fibers in
the bundle is

packing fraction (hexagonal close-pack)

$$= \left(\frac{N}{n_d^2}\right)\left(\frac{d_{core}}{d_{clad}}\right)^2 \qquad (7.1)$$

where d_{core} is the core diameter of the fiber and d_{clad} is
the total fiber diameter (core + clad). Clearly, the
highest packing fraction is achieved using fibers with
inherently very thin cladding or fibers where the cladding
is essentially stripped from the core over the short length
of the coupler. When $d_{core} \simeq d_{clad}$, the packing fraction
loss becomes (N/n_d^2), which is approximately -1.2 dB for
hexagonal bundles containing either 7, 19, 37, or 61 fibers.
For bundles assembled using fibers with a thick cladding
which has not been stripped from the core, such as fibers
with 125 μm overall diameter and 75 μm core diameter, the
packing fraction loss increases by the factor $(d_{core}/d_{clad})^2$
which equals -4.4 dB for the diameters mentioned above.
For this case, the total bundle packing fraction loss is
-5.6 dB.

The packing fraction loss for a linear, close-packed
bundle, again containing a total of N fibers, is

packing fraction (linear close-pack)

$$= \frac{\pi}{4}\left(\frac{N}{1 + (N-1)\dfrac{d_{clad}}{d_{core}}}\right) \qquad (7.2)$$

The highest packing fraction is achieved using fibers with
thin cladding. For the linear bundle with $d_{clad} \simeq d_{core}$,
the packing fraction loss becomes $\pi/4(-1.05$ dB), independent of the number of fibers. For bundles assembled with
fibers with 125 μm outside diameter and 75 μm core diameter,
the packing fraction loss increases by a factor of

**19 STRAND HEXAGONAL
CLOSE PACKED BUNDLE**

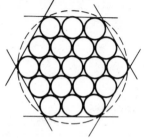

**7 STRAND HEXAGONAL
CLOSE PACKED BUNDLE**

-1.1 dB
PACKING FRACTION LOSS
(WITH FIBER CLADDING
REMOVED)

-1.2 dB
PACKING FRACTION LOSS
(WITH FIBER CLADDING
REMOVED)

**6 STRAND LINEAR
CLOSE PACKED
BUNDLE**

RESULTS ARE
INDEPENDENT
OF FIBER SIZE

-1.05 dB
PACKING FRACTION LOSS
(WITH FIBER CLADDING
REMOVED)

Fig. 7.4.   Basic bundle configurations for close-packed
fiber bundles.

approximately -2 dB for bundles containing six to 20 fiber waveguides. The total packing fraction loss for these bundles is, therefore, approximately -3 dB, which is approximately 2.5 dB better than the loss associated with the hexagonal format. It is interesting to note that when the cladding is thin, the packing fraction loss is approximately -1 dB, both independent of the number of fibers used and of the bundle format (hexagonal or linear). If the clad is relatively thick, as is currently the case for Corning low-loss optical waveguides, the packing fraction loss for a flat, linear ribbon geometry bundle is approximately 2.5 dB better than that of the hexagonal bundle.

Since the packing fraction loss is a function of cladding thickness, the above computations reveal that the packing fraction loss can be substantially reduced by arranging the fibers in a close-packed configuration and by using fibers with thin clad thickness in the coupler region. The latter can, of course, be achieved by using fibers which are manufactured with thin cladding or by removing a substantial portion of cladding material from fibers manufactured with relatively thick cladding. If thick-clad fibers are used, the initial temptation is to remove, by controlled etching, essentially all the cladding material from the relatively short length of fiber which is encased in the input connector. If, however, the fibers in the input connectors are held fixed in the desired close-packed configuration using relatively absorbing epoxy material, a considerable amount of optical loss can occur in the short, clad-stripped section of fiber if too much of the cladding is removed.

Hudson and Thiel[5] have experimentally determined that a minimum in the single fiber packing fraction loss occurs when the cladding thickness is approximately 5 µm. For this thickness of clad, the measured packing fraction loss of a <u>single</u> 85 µm core diameter fiber, surrounded by epoxy, is -1.5 dB. The total length of the thinned clad portion of the fiber was 3 cm. The experimental results described in Ref. 5 indicate that an improvement in the packing fraction loss of thickly clad, low-loss fibers is possible. The connector designer must exercise caution in the amount of cladding material removed. The correct amount to remove will be a function of both the length of the thinned clad fibers and of the material in which they are embedded.

6. INPUT CONNECTORS

Fiber bundle input couplers with insertion losses
equalling the bundle packing fraction losses, in principle,
can be fabricated if the source size and bundle format are
judiciously chosen. The preceding computations indicated
that the best achievable coupler insertion loss (neglecting
Fresnel losses which are on the order of -0.5 dB if the
fiber ends are not AR-coated) is approximately -1 dB. The
realization of this low value of input coupler insertion
loss depends greatly on the development of high brightness
(efficient, small area) light emitters, either lasers or
LED's, and the engineering development of precision couplers
housing the source and lens. These couplers could take the
form of a "hard-wired, pigtailed" unit containing a short
pigtail section of fiber bundle, a lens system, and a source,
all aligned anf fixed permanently into position as a single
coupler unit. The opposite end of the short pigtailed
section of fiber bundle could then be affixed with a quick
connect-disconnect splice coupler for coupling to the
fiber cable in the system. An alternative to the "hard-
wired, pigtailed" approach is to provide the quick connect-
disconnect function in the image plane of the lens itself.
As a result of the alignment tolerances required between
the lens and the fiber bundle, this approach may not be
economically as attractive as the pigtailed approach. In
either case, since the source and bundle format are critical, for a precision coupler to be of practical value,
some fiber system component standardization should first
occur.

In the short term, more readily interchangeable, less
mechanically precise, higher insertion loss couplers can
be fabricated by merely butting the source directly to the
fiber bundle. This will result in the numerical aperture
loss of $(n+1)/2NA^2$, the packing fraction loss, and any area
mismatch loss which results if the source emission size is
larger than the fiber bundle. The area mismatch can be
minimized by judicious choice of source size and bundle
format. The numerical aperture loss is, of course, less
severe for higher NA fiber. For example, several manu-
facturers are currently producing prototype components
which include receiver and transmitter terminals connected
to relatively high NA (NA > 0.4), high packing fraction
fiber bundles by directly butting the ends of the fiber

214

bundle against the light-emitting diode in the transmitter and photodetector in the receiver. For a fiber bundle having an NA of 0.5 and a packing fraction loss of -1 dB, the input coupling loss for a Lambertian LED is -7 dB, a tolerable value in many applications.

7.   ALLOWABLE DISTRIBUTION SYSTEM LOSSES

In order to estimate how much optical power is available for distribution of the data, compare a future system wired first with 350 dB/km cable, then 100 dB/km cable, and finally with 5 dB/km cable. Also, assume that the sources and bundle configurations have been carefully selected so that the only input coupling losses are those caused by the packing fraction of the bundle. Typical fiber bundles with 350 dB/km attenuation losses which are currently available are comprised of fibers whose outside diameter is ∿46 μm with a ∿43 μm core diameter. The numerical aperture of this high-loss fiber is 0.63. The total diameter of the fiber bundle depends on the number of fibers used. Bundles containing 200 fibers are approximately 1.27 mm in diameter, while a bundle with ∿5000 fibers is approximately 3.2 mm in diameter. The currently available medium-loss (100 dB/km) fiber bundles contain approximately 60 strands of 65 μm outside diameter and 55 μm core diameter fiber whose numerical aperture is 0.5. The diameter of the bundle is approximately 0.56 mm. Finally, the low insertion loss (5 dB/km) fiber cables comprised of fibers with 125 μm outside diameter and ∿75 μm core diameter with an NA of ∿0.14 are available in both seven- or 19-strand hexagonal format bundles. The entrance diameter of these bundles with the fibers completely clad is 0.38 mm and 0.63 mm, respectively. If the cladding is removed to improve the packing fraction, the entrance diameter becomes 0.23 mm and 0.38 mm for the seven- and 19-strand cables, respectively.

For the large diameter, high-loss fiber, a large emission area LED such as, for example, the Texas Instruments SL1314 device (see Fig. 7.3), capable of emitting +10 dBm of optical power when driven at ∿300 mA, is a suitable choice for the source. This particular device is a 0.46 mm diameter dome emitter packaged in a parabolic reflector so that the radiation emitted from the packaged unit is contained within ±20°. Since the emission aperture of the

unit, which is 2 mm in diameter, is also less than the
bundle area, all the optical power emitted from the device,
less packing fraction and reflection loss, is coupled into
the fiber transmission line. Since the packing fraction
loss and reflection loss for the large diameter, high-loss
fiber bundle are approximately -1.0 dB, the optical power
at the input to the transmission line is +9 dBm for the
example chosen.

Consider now the medium-loss (100 dB/km) fiber cable
and the low-loss (5 dB/km) fiber cable. Since in both
cases the cable diameter is on the order of 0.5 mm or less,
a high brightness emitter (such as a Burrus-type LED or an
edge-emitting LED or laser) and an intervening optical
system is selected so that the only input coupling insertion
loss is that caused by packing fraction and reflection loss.
Since the best achievable packing fraction loss for these
bundles is approximately -1 dB, independent of bundle
configuration, the best that could be expected in total
coupler insertion loss is -1.5 dB.

Reference to Fig. 7.3 reveals that with approximately
150 to 250 mA of drive current (depending on the particular
device chosen) the high brightness emitters are capable of
emitting approximately +4 dBm of optical power. For both
the medium- loss and the low-loss cables the optical power
at the input to the transmission line is, therefore,
approximately +2.5 dBm.

Having established the optical power coupled into the
transmission line for the three different cable types, the
total amount of optical power that can be expended in data
distribution to remote terminals can be determined by
plotting, as in Fig. 7.4, the optical signal level in the
transmission line as a function of the terminal-to-terminal
spacing. Using, for the purpose of calculation, a receiver
sensitivity limit of -55 dBm and a maximum terminal spacing
of 100 m, the plot in Fig. 7.5 indicates that the distribution system loss can be -29 dB for a system wired with
350 dB/km cable, -48 dB if 100 dB/km cable is used, and
-58 dB if 5 dB/km cable is employed. Recall that at 40
to 50 Mbit/sec data rate and a 10^{-9} error rate, the minimum
detectable power required using a PIN photodiode is -50 dBm,
while that with an APD is -60 dBm.

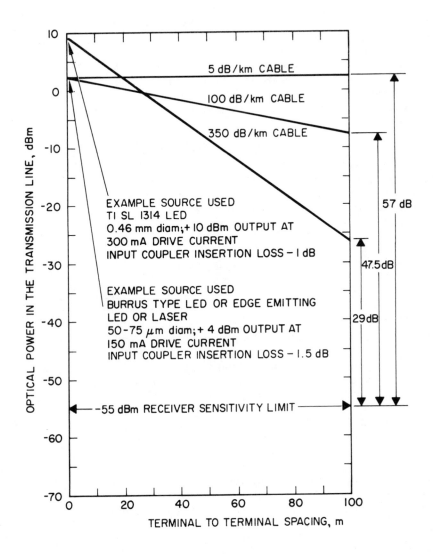

Fig. 7.5. Optical signal level in transmission line as a function of terminal-to-terminal spacing.

8. DISTRIBUTION SYSTEMS

There are currently two configurations being considered for the distribution of data to a set of remote terminals.[8,9] One is a serial distribution system which employs access couplers, and the other is a parallel system employing a "star" coupler. A schematic diagram of an N-terminal distribution system with access couplers is shown in Fig. 7.6; the schematic for a similar N-terminal parallel system is shown in Fig. 7.7.

Hudson and Thiel[5] have shown that for a serial distribution system with access couplers with a constant tap ratio, the lowest ratio of optical power in the transmission line at the input to one terminal to the optical power at the output to another occurs between terminals 1 and N-1 in an N terminal system. The ratio is

$$\frac{P_{N-1}}{P_1} = (2L_C + L_{CI} + L_{IT})\ (N-3) + (2L_C + L_{CI} + L_T) + L_S$$

$$(7.3)$$

where

L_{CI} = internal insertion loss of the access coupler

L_C = insertion loss of the cable couplers attached to the access coupler at each of the three ports

L_S = splitting factor of the duplex input-output coupler necessary for bidirectional operation. This always has the value -3 dB.

L_T = tap ratio of the access coupler

L_{IT} = insertion loss associated with power tapped by the coupler.

The above ratio assumes that the access coupler is symmetric, that is, the insertion loss is constant, independent of which pair of access coupler ports is being considered. On the other hand, the corresponding ratio of optical powers in a parallel system employing a star coupler is independent of which pair (jk) of system terminals is being considered and is given by[5]

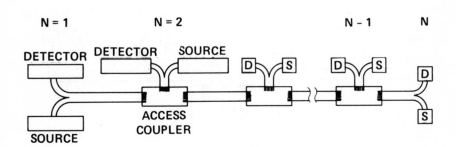

WORST CASE POWER RATIO

$$\frac{P_{N-1}}{P_1} = (2L_C + L_{CI} + L_{IT})(N-3) + (2L_C + L_{CI} + L_T) + L_S$$

$$L_S = \text{SPLITTING FACTOR} = 3 \text{ dB}$$

$$L_T = -10 \text{ LOG}(1 - \text{TAP RATIO}) = 0.46$$
$$\text{(FOR CONSTANT 10 dB TAP RATIO)}$$

Fig. 7.6   Schematic diagram of an N-terminal serial
distribution system with access couplers.

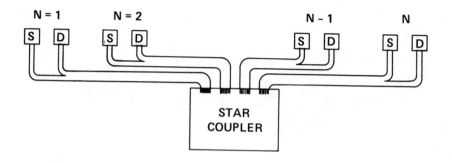

ALL TERMINALS

$$P_j/P_k = 4L_C + L_{CI} + L_T + L_S$$

$$L_T = -10 \text{ LOG } 1/N$$

$$L_S = \text{SPLITTING FACTOR} = 3 \text{ dB}$$

Fig. 7.7. Schematic diagram of an N-terminal parallel distribution system with star coupler.

$$P_j/P_k = 4L_C + L_{CI} + L_T + L_S$$

where

L_C = the insertion loss associated with the cable connectors

L_{CI} = insertion loss of the star coupler

L_S = splitting factor of bidirectional input-output connector

L_T = 10 log(1/N) is the tap ratio or splitting factor of the star coupler.

A comparison of the parallel and serial systems can be made by plotting distribution system losses versus the number of terminals for the two systems. Such a plot is shown in Fig. 7.8 for a cable connector insertion loss of 1 dB for both the parallel and the serial system, an access coupler insertion loss of 2 dB with a constant 10 dB tap ratio, a star coupler insertion loss of 7 dB and, since the system is assumed bidirectional, a 3 dB splitting factor. Thus, for a system with a maximum interterminal separation of 100 m and a receiver sensitivity limit of −55 dBm, the maximum number of terminals that can be accessed varies from ∿6 to 12 (depending upon the cable loss) for the serial system and from ∿32 to 20,000 for the star distribution system. These results, of course, are dependent on the insertion losses of both the star and the access couplers. In addition, it is very much dependent on being able to achieve the −55 dBm receiver sensitivity limit used in this example.

The plots shown in Fig. 7.8 clearly illustrate the signal level advantage of the star system over the serial system for even a system with a limited number of terminals (for example, 10). In addition, the receiver in the serial system must be equipped with a wide dynamic range AGC to handle the strong signals from adjacent terminals and the weak signals from remote terminals. Since the parallel system has but a single mixer, it does not have this dynamic range problem. The added uniform signal level available with the parallel system translates to less stringent design requirements on both the transmitters and the receivers. The cost of this added signal level is paid out in the amount of fiber cable necessary to wire the system. The

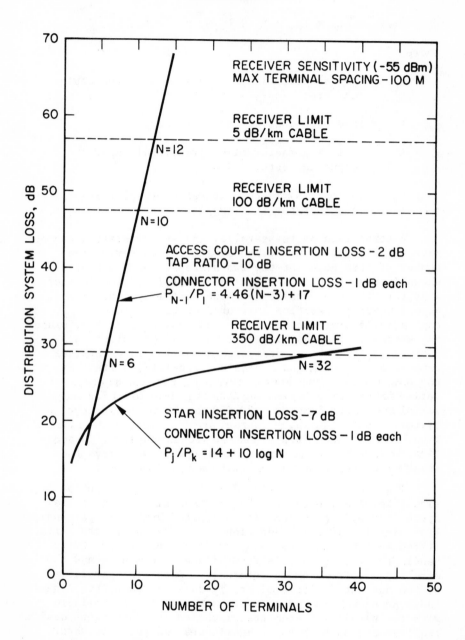

Fig. 7.8.   Distribution system loss plotted as a function
of the number of terminals.

star bus design in essence shortens the main bus to a
single point mixer and extends the length of each terminal
arm.

Distribution System Components. Seven terminal star
couplers with average insertion losses of 7 dB have been
fabricated[5] and incorporated in a flight test demonstration
of an actual distribution system.[6,7] The fiber optic data
bus was used for carrying flight control signals inside a
modified Air Force C-131 aircraft. The flight test was
associated with a multiplex program whose goal is to
develop a multiplexed data transmission system for inter-
connecting avionics. The overall objectives of the program
are to reduce wiring weight and volume and to increase
system flexibility and adaptability. During the initial
phase of the program, a three-terminal redundant system for
carrying multiplex flight control signals was developed
and flight tested using twisted wire pair, microwave waveguide, and fiber optic transmission lines. Details of the
flight test can be found in the contract report listed in
Ref. 11.

The fiber optical waveguide data bus consisted of a
star coupler and seven fiber bundles, each containing 61
strands of multimode, low-loss glass fiber. The system,
which was configured as shown in Fig. 7.9, consisted of
three terminals which multiplexed and demultiplexed the 16
analog electrical flight control signals of the fly-by wire
flight control system.

In the flight test experiment conducted, the maximum
terminal-to-terminal spacing was approximately 100 ft.
Since each terminal in the system was required to monitor
its own transmission, the maximum optical path length
traversed by an optical signal was 200 ft. Two fiber
bundles were terminated in each terminal, one for the
transmitter and one for the receiver. Commercially available
(the diodes used were Texas Instruments SL1314, whose power
output versus drive current are plotted in Fig. 7.3) GaAs
light-emitter diodes were employed as the optical source,
while commercially available silicon PIN photodetector-
amplifier modules were used in the receiver. Input and
output connectors were fabricated using modified BNC connector bodies. The nominal optical output power incident
on the receiver photodiodes was −28 dBm. Although the

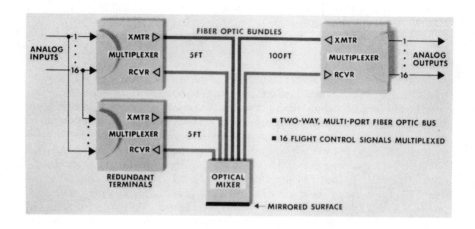

Fig. 7.9. System layout of flight-tested star data
distribution system.

optical bus was designed with a 10 mHz bandwidth, the
information transfer rate which was determined by other
system considerations was 1/2 Mbit/sec for all three transmission systems evaluated. The optical bus operated for
the full duration of the flight test (40 minutes) without
any detectable errors. The successful results of this
first flight test help demonstrate the feasibility of
utilizing fiber optic transmission lines in avionic systems.

There are several basic designs of serial access
couplers being considered for distribution systems using
fiber bundles. In the design being pursued[8] at the Naval
Electronics Laboratory Center (NELC), the coupler consists
of a glass block with internal, fully reflecting mirrors
which serve to deflect the light signals into or out of the
main mixing block. A schematic illustration of a bidirectional coupler of this type is shown in Fig. 7.10. The
tap ratio for this type of T-coupler depends on the size
of the internal mirrors. Unidirectional couplers, couplers

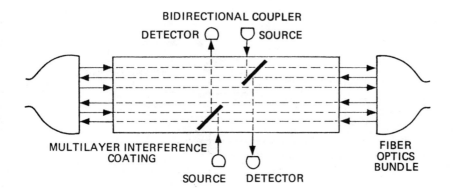

Fig. 7.10. Bidirectional access coupler using glass
mixing block with internal mirrors.

with only a single internal mirror, have been fabricated
with a total throughput loss[8] of -4.8 dB.

T-access couplers of the type shown schematically in
Fig. 7.11 formed by bonding bent Pyrex rods to a mixing
rod are being considered[9] at the Naval Research Laboratories (NRL). The unidirectional coupler illustrated in Fig.
7.11 has been fabricated and found to have an insertion loss
of ~1.5 dB, exclusive of the tap loss. The loss was measured on the Pyrex coupler structure which was held by felt
pads and illuminated with a fiber bundle using an incoherent
source. It remains to determine the effects of incorporating
such a structure into a practical connector package.

The third type of access coupler uses bifurcated fiber
bundles in conjunction with glass mixing rods. Couplers
of this variety have recently been fabricated[10] with an
internal insertion loss of 1.5 dB.

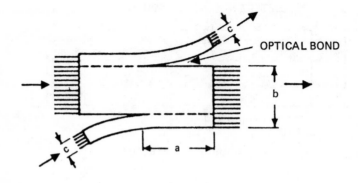

Fig. 7.11.   T-access coupler fabricated using bent Pyrex
arms bonded to a mixing rod.

9.   SUMMARY

In this chapter an elementary treatment of some of the
engineering design considerations that must be addressed
in designing data distribution systems using fiber optic
transmission lines have been presented.   A comparison of
serial and parallel distribution systems has been made.

REFERENCES

1. M. DiDomenico, Jr., Industrial Research, p. 50, August 1974.

2. C. M. Stickley et al., "Anticipated Uses of Fiber Optics and Integrated Optics in the Defense Department," presented at Topical Meeting on Integrated Optics, New Yorleans, La., January 21-24, 1974.

3. I. Hayashi, Applied Physics 5, 25 (1974).

4. C. A. Burrus, Proc. IEEE 60, 231 (1972).

5. M. C. Hudson and F. L. Thiel, Applied Optics 13, 2540 (1974).

6. J. D. Anderson, M. K. Barnoski and A. S. DeThomas, "Fiber Optic Data Bus," presented at Integrated Optics and Fiber Optics Communications Conference, NELC, San Diego, Calif., May 15-17, 1974.

7. "Fault-Tolerand Digital Airborne Data System Flight Test," Technical Report AFFDL-TR-74-122, December 1974. Air Force Flight Dynamics Laboratory, Wright-Patterson Air Force Base, Ohio.

8. H. F. Taylor et al., "Fiber Optics Data Bus System," NELC Report No. TR 1930, August 26, 1974.

9. A. F. Milton and A. B. Lee, "Access Couplers for Multi-terminal Fiber Optic Data Communication Systems," Technical Digest of Topical Meeting on Optical Fiber Transmission, Williamsburg, Va., January 7-9, 1975.

10. Dr. R. Baird, Spectronics, Inc., private communication.

A 6
B 7
C 8
D 9
E 0
F 1
G 2
H 3
I 4
J 5